DATE			

Solutions to Boiler and Cooling Water Problems

Solutions to Boiler and Cooling Water Problems

SECOND EDITION

Charles D. Schroeder

VNR VAN NOSTRAND REINHOLD
New York

Copyright © 1991 by Van Nostrand Reinhold

Library of Congress Catalog Number
ISBN 0-442-00501-6

Printed in the United States of America

Van Nostrand Reinhold
115 Fifth Avenue
New York, New York 10003

Van Nostrand Reinhold International Company Limited
11 New Fetter Lane
London EC4P 4EE, England

Van Nostrand Reinhold
102 Dodds Street
South Melbourne 3205, Victoria, Australia

Nelson Canada
1120 Birchmount Road
Scarborough,Ontario MlK 5G4, Canada

16 15 14 13 12 11 10 9 8 7 6 5 4 3 2 1

Library of Congress Cataloging in Publication Data

Schroeder, C. D.
  Solutions to boiler and cooling water problems/Charles D. Schroeder.—2nd ed.
  p.   cm.
  Includes bibliographical references and index.
  ISBN 0-442-00501-6
  1. Steam boilers—Maintenance and repair.   2. Water supply, Industrial.   3. Feed water—Purification.   I. Title. TJ289.S38   1991
621.1'83—dc20   90-12607
                                                                         CIP

Contents

Preface

With the emphasis on the environment in recent years, rapid changes are taking place in the treatment of boiler and cooling water systems. Changes have not only included the addition of new and improved products but the elimination of long established water treatment chemicals. Unfortunately, some of the older products were superior in performance. As a result of these changes, new and additional problems are developing that require solutions.

Two major products that have been eliminated are chromates and pentachlorophenates. Chromates, alone or combined, provided excellent corrosion inhibition in the majority of installations. They had become the standard against which other inhibitors were compared. In addition, in many cases chromates prevented the development of sulfate-reducing bacteria. SRB is a major factor in microbiological corrosion.

Sodium pentachlorophenate was probably the most effective biocide available for industrial cooling water systems. Unfortunately, both chromates and sodium pentachlorophenate were too toxic for the environment. This has resulted in their virtual elimination.

Their elimination has not been without a price. It is more difficult to protect systems against corrosion and slime formation. As a result, different and more numerous treatment problems have developed in recent years.

After many years during which ion-exchange softener represented the vast majority of pretreatment installations, we find other methods making their appearance. Of greatest importance is reverse osmosis.

In view of the new developments in chemicals and equipment, this second edition of *Solutions to Boiler and Cooling Water Problems* has been prepared. Also, selected changes have been made in the description of material and other water treatment chemicals have been added. Essentially, the method of covering problems and solutions remains the same as in the original edition. Also as in the first edition, the use of case histories has been virtually ignored. Instead, steps are suggested that assist in identifying a problem; usually, the cause of a problem points the way to a solution.

As an example, pitting in a cooling water piping system might automatically indicate to some operators that oxygen pitting had taken place. However, if a black deposit at the base of a pit is nonmagnetic, the evidence is strong that the pit was caused by sulfate-reducing bacteria. Other suggested tests will confirm this diagnosis.

Often, this book defines a problem and proposes a number of possible solutions. The operator must then make the final decision on the appropriate action to take. This would occur, for instance, with the impingement (grooving) corrosion of copper tubing. Several options are available to solve this problem. The copper can be replaced with another metal, velocity reduced directly, temperatures reduced, tubing diameter increased, sharp elbows reduced, etc. After all the pertinent factors are reviewed, an informed decision can be made on the steps required to solve this problem in a particular plant. The outlined method of troubleshooting is also covered in the "How to Use This Book" section of the introduction.

This book is directed toward anyone involved in the operation of boiler and cooling water systems. This includes staff engineers, plant engineers, plant chemists, and plant managers. Subject matter relates only to those problems likely to occur in such plants and to their possible solution. It is hoped that this book will enable plant personnel to begin diagnosis of water problems as soon as symptoms are noticed, thereby limiting equipment damage and other losses. This book will also help readers become more effective users of services provided by water treatment companies and independent consultants.

Although the water treatment industry has grown and matured over the years, its approach to problems has often been disorganized. Too often, the person in charge of water treatment has had to rely on his own experience exclusively or, lacking that, his ingenuity. At times, this approach succeeds. If failure results, a different approach is tried, often with no logical reason for its selection. This book has been written in the hope that industrial water treatment problems can be approached in a more systematic manner. It fills a gap in the water treatment literature and should be helpful to many individuals working in water treatment and related fields.

Water often acts in unpredictable ways. The vast majority of water treatment books contain a description of water treatment methods involving equipment and/or water chemistry. Even when a few problems are included, the main purpose of these books is not problem solving. To obtain answers to the many problems that arise would involve maintaining a library larger than can be expected in plant offices.

Other than water treatment books, the main sources of information are periodicals. Many of these publications cover new developments. While interesting and valuable, they usually do not offer the operating engineer much help in solving daily problems.

In some cases, this book will provide the reader with new answers. Often, the solutions presented are familiar to the engineer or chemist, but this book will serve as a reminder and as assurance that all recognized solutions are included in each problem analysis. By restricting the subject matter to problem solving it is hoped that the person in charge of boiler and cooling water operations will be able to find many of the answers to the numerous water problems that develop during the life of a plant.

The solutions proposed usually are directed toward the plant in its existing condition. A plant may be using hard-water makeup to a boiler that requires a high percentage of makeup.

Obviously, this is an undesirable situation, and a softener should be installed. For various reasons, the plant engineer may have to operate the plant as it is—without a softener. Accordingly, most solutions are directed to improving operations without major equipment changes, although long-term solutions are also suggested.

Water treatment is such a broad field that restrictions in coverage should be stated. Primarily, only in-plant operations are considered. This includes pretreatment plus the boiler and cooling water systems. Municipal water treatment systems, wells, and waste-water problems are beyond the scope of this book. Waste heat boilers are also excluded since a large number of these units are of special design.

This book focuses on fossil-fueled steam boilers because of their wide use in favor of high-temperature hot-water systems. In covering this subject, it is assumed that the reader is acquainted with industrial water treatment problems. Accordingly, only brief coverage is given to descriptions of water treatment processes, equipment, etc. Many fine technical books are available on the various aspects of the subject for readers who desire additional information. No one book can possibly cover all the water treatment problems that arise. If a direct answer cannot be found here, it is hoped that the reader can be led in the right direction. A brief description of various processes is included to aid the reader in reviewing the subject matter and as a context for solving problems.

Of the many problems that arise in a boiler or cooling water system, some are purely electrical or mechanical; they are dealt with in detail in specialized literature. Others clearly fall into the domain of water treatment. There are, however, many areas that are not clearly defined. Probably, the majority of plant water problems involve both water chemistry and physical factors. For example, water may be present in a heat exchanger tube, but failure may result purely from fatigue. Or a boiler tube may fail from overheating, which, in turn, may relate to scale. Accordingly, problems in this book will cover areas in which water (liquid or gas) is present. Water chemistry may or may not be a part of the problem. Corrective measures may involve water chemistry or physical factors; changes in both may be required.

The type of problem encountered changes with the equipment system considered. In a boiler or cooling water heat exchanger, the major emphasis is on preventing corrosion and deposits. On the other hand, in an ion-exchange softener or deaerator, the majority of problems relate to performance. Examples are excess hardness in the effluent of a softener or high oxygen levels in the effluent of a deaerator.

The major objective of this book is to cover real and potential problems and their solutions and to offer the reader a systematic approach to diagnosis and corrective action. Thanks are given to the various companies that provided photos or sketches. Special thanks are given for the support of my wife, Betty Doris; also my daughters, Andrea, Janet, and Suzanne, who put in many hours of typing and proofreading.

C. D. Schroeder
Sioux Falls, South Dakota

Introduction

General Considerations

Whenever there is a departure from normal water treatment operation, the possibility of a problem developing is present. If the phosphate reading in boiler water suddenly drops or if sludge suddenly appears in boiler water that was previously clear, an investigation should be made.

In many cases, the answers to problems are obvious, so that all simple solutions should be checked first. In the above case where phosphate levels fell, for instance, an operator may simply have failed to add phosphate. Where sludge developed in the boiler water, perhaps a softener was not regenerated. The first step should be to establish that normal plant operations are being followed. When it has been determined that all recommended operations are in order, then it is time to make a more thorough investigation.

In the above cases, the investigator who ignored this practical procedure might needlessly send a boiler water sample to a central laboratory or return the water treatment chemical to a supplier. In short, if a problem is simple, it should not be made complicated.

Personnel

If normal operating conditions are apparently being followed and unusual results are still obtained, the problem warrants further investigation. During fact-finding, the investigator should bear in mind that the competence of workers varies and should double-check that the verbal information he receives is correct. It is not enough to be informed that an ion-exchange

1

softener is delivering soft water. If at all possible, the investigator should make his own determination.

Conversely, competent workers can be a valuable source of information. Their acumen often results in their knowing the cause of a problem. Accordingly, good rapport with them is a necessity in providing solutions.

Data

The plant that maintains the best records is also the plant that is best able to solve any water treatment problems that develop. It is difficult to determine if a plant has deviated from normal operation if no records are maintained. Records not only provide background for solutions to immediate problems but may also indicate trends that can lead to problems. As an example, increases in an ion-exchange softener rinse time may indicate resin deterioration. Basic, but important, data should include the amounts of water treatment chemicals used on a daily basis. If the amount of phosphate required increases, it is a good indication that the feedwater hardness has increased or the boiler load has increased. If a larger amount of sodium sulfite is required, it usually indicates problems in the deaerator, etc.

History

As previously noted, the simple solution to a problem should be investigated first. For example, a plant may have experienced boiler deposits resulting from carryover from a hot-lime treater in past operations. If a new deposit problem develops, the obvious first step should be to investigate the lime softener again. Only after the hot-lime treater has been cleared should the other paths of investigation be taken.

Plant Age

The age of a plant can have an influence on the type of problems encountered. Mistakes can be made in the construction of a new plant. It is quite common for errors to be made in the choice of materials of construction: mild steel might be employed where stainless steel should have been specified. Or a pipefitter might have installed a horizontal check valve in a vertical line. Such mistakes may be compared to the shakedown voyage of a new ship. In time, these errors are uncovered and corrective steps are taken. Keeping the above comments in mind, the problem solver should know the approximate age of a plant. If the plant is new, errors in construction must be considered. In older plants, these factors no longer carry the same importance.

On the other hand, certain problems can be expected to develop in older plants. For example, ion-exchange resins eventually deteriorate, sections of pipe eventually corrode, etc.

Analytical Results

Quite often, operators feel that they face a possible problem because abnormal results are obtained in their water treatment testing procedures. As an example, in a plant the author investigated, the boiler operator stated he could not obtain any alkalinity in his boiler. This condition existed even though additional caustic had been added to the boiler water. Investigation revealed that the operator was using concentrated sulfuric acid in his tests for alkalinity instead of the specified N/50 acid.

When erratic test results are obtained, the test reagents and test methods should first be checked. After results are verified, further investigation can then be made.

If a consulting water treatment company is employed, results obtained using the consultant's reagents should be crosschecked against results obtained using plant reagents.

How to Use This Book

The methods used in this book to identify water treatment problems can be compared with the approach of an inexperienced tree buff to the problem of identifying a particular tree. The tree is a scarlet oak, but he doesn't know this. To identify the tree, he could search through a tree book containing 500–600 illustrations and descriptions of individual trees. Needless to say, this procedure is very time-consuming, and few people would use this method. Instead, attempts are first made to classify before making a final identification.

In the example above, the investigator would probably note whether the tree was a conifer (cone-bearing) or broad-leafed. Since an oak tree is broad-leafed, all cone-bearing trees are eliminated from the search. Next, it might be noted whether the leaf was a single-leaf or part of a compound-leaf. Since oaks are single-leaf, all compound-leaf trees (chestnut, ash, hickory, etc.) are eliminated.

The tree book reveals that most oak leaves have a characteristic shape. This eliminates the maples, sycamores, etc., which have leaves of a different shape. The oaks can then be further divided into the white and red oak groups. White oaks typically have rounded lobes while those of the red oaks are pointed. The scarlet oak happens to be a type of red oak.

If a few of the red oaks have similar leaves, the size and shape of the acorns can be compared.

Using the above method of classification, only five or six checks have to be made to obtain a positive identification. The investigator is constantly trying to determine if certain known characteristics match the tree he is examining.

In a similar manner, this book first tries to identify types of water-related problems. When the problem is identified, an examination of possible causes is made. Knowing the cause of a problem, a solution is then proposed.

To illustrate the method, a boiler inspection might disclose heavy deposits in the tubes; in the past, the tubes had been clean. Basically, new boiler deposits can originate from three major sources. First, the feedwater quality may have changed; second, the internal treatment may have changed; third, the physical operating conditions may be undesirable.

The table of contents is the key to identifying and providing a solution to problems. In the above case, the three factors can be examined under Part II, Chapter 5. In this example, it might develop that the feedwater quality or the physical operating conditions have not changed. These two classifications are then discarded.

A check of the internal chemical treatment might disclose that dosages were low or that chemical controls were maintained in the wrong range. It should be noted that eight types of typical internal treatments are listed. No plant will employ all eight methods. Accordingly, only the comments under the treatments being used should be considered. Possible problems in the use of these methods are covered; corrective measures are suggested. Occasionally, more than one operating condition must be corrected to provide a solution.

Two different types of problem conditions can exist. In one, a boiler may have operated for years with no scale problems. Then suddenly, deposits develop. In attempting to correct this problem, changes in operating conditions should be noted. Recordkeeping is important.

A more difficult type of problem occurs where a boiler has never operated without scale. Under these conditions, problem solving can be more difficult. Standard operating conditions have not prevented scale formation. In this case, a change in internal treatment may have to be made or blowdown increased. In short, any steps that can be taken to decrease scale should be considered.

In some cases, changes in treatment methods are not sufficient to solve a problem. If the silica, for example, in the feedwater of a high-pressure boiler is very high, only the installation of proper pretreatment to remove silica will solve the problem.

Raw-Water Quality

In most products, the quality of the raw materials used is a determining factor in the quality of the finished product. The same can be said of water; but the analogy is limited. The water treatment industry has developed around making raw water suitable for its intended end use. Still, many water treatment problems originate with poor-quality raw water; other problems are caused by changes in raw-water composition.

Needless to say, a knowledge of the properties of the raw water employed is of major importance in solving water treatment problems. In comparing raw-water analyses, it is important that the items being studied are expressed in the same units. It has become customary in water treatment to express analyses in several different forms. Chlorides, for example, can be expressed as ppm Cl, ppm $CaCO_3$, epm Cl, and grains per gal Cl. Parts per million (ppm) and ppm $CaCO_3$ are in the widest use.

In alphabetical order, the items of importance in reviewing raw-water quality are as follows.

Alkalinity (M)—Total.   This unit is not a measure of a specific ion but the total alkalinity. Primarily, total alkalinity in water covers bicarbonate, carbonate, and hydroxide. Also included, but of lesser importance, are phosphate and silicates. In raw water, total alkalinity primarily covers bicarbonate. A small amount of carbonate might also be present, while little or no hydroxide is available. Total alkalinity in raw water is important since it represents a potential source of scale. It also indicates the amount of acid required to neutralize the water to the neutral range. Specific properties are covered under the individual ion headings. M alkalinity corresponds to an endpoint pH of 4.3.

Alkalinity (P) Phenophthalein.   This unit is used to cover all OH alkalinity plus one-half the carbonate alkalinity. Most often, it is not present in raw water. This corresponds to an endpoint pH of 8.3.

Ammonia.   In raw water, ammonia can originate from fertilizer runoff or pollution. Bacteria usually convert the ammonia in raw water to nitrite and nitrate. In most circumstances, little or no ammonia is present in raw water. Specifically, ammonia is corrosive to copper and copper alloys.

Bicarbonate-Carbonate.   As previously mentioned, the primary source of alkalinity in raw water is the bicarbonate ion. A small amount of carbonate may also be present. Upon heating, the bicarbonate ion decomposes to carbonate and carbon dioxide. If a soluble salt results, further heating decomposes the carbonate to caustic (OH) and carbon dioxide.

Caustic in boiler water may or may not be desirable, depending on the operating pressure. Carbon dioxide is a primary source of condensate line corrosion.

Calcium.   In the vast majority of raw waters, calcium is the main source of hardness. In boiler and cooling water systems, calcium salts are the main cause of scale. Overall, calcium is found in larger quantities in ground water vs. surface water. Although calcium is primarily known for its scale-forming tendencies, it also acts as a cathodic corrosion inhibitor. All other factors being equal, a water containing calcium is not as corrosive as a soft water.

Chloride.   Scale is not a problem where chloride salts are concerned, but chloride salts are very corrosive in an oxidizing environment. Chlorides are present in virtually all raw waters. In analyzing corrosion problems, the chloride content should be noted. Chlorides are prominent in crevice corrosion and pitting. Austenitic stainless steels are subject to pitting, crevice corrosion, stress corrosion cracking, and intergranular corrosion from chlorides. In all cases of design or operation, steps should be taken to prevent the concentration of chlorides.

Iron.   For the most part, soluble iron is found in ground water although complexes can also be found in surface waters. In ground water, iron is normally found in the soluble ferrous form. On contact with air or other oxidizing agents, the iron is converted to the ferric form. In this state, it forms insoluble hydroxide or oxides. If the iron content of the raw water is 0.3 ppm or greater, it should be removed or controlled. Indirectly, iron contributes to corrosion by forming deposits. Differential aeration cells form under the deposits. Iron-bearing waters also contribute to the growth of iron bacteria and sulfate-reducing bacteria.

Magnesium.   This cation is the other main component of water hardness. In most cases, it is found in lesser amounts than calcium. Magnesium usually forms hydroxides or silicates in boiler water. They are desirable types of sludge since they are readily dispersed. If the boiler alkalinity is low, undesirable magnesium phosphate often forms. This is sticky and scale-forming. In cooling water systems, magnesium precipitates only when the pH is over 10.0.

Manganese.   Although not present as frequently as iron, this cation is often found in iron-containing waters. It has many of the same properties as iron and should be controlled or re-moved from raw water if the content exceeds 0.2 ppm. The color of manganese deposits is dark gray or black. It is more difficult to oxidize than iron; generally, a high pH is required for its removal by precipitation, in addition to an oxidizing environment.

Nitrate.   Industrially, the presence of nitrates in raw water is not a problem. However, its presence often indicates that the raw water is, or has been, polluted. Another reason for the presence of nitrate can be runoff in the use of fertilizer.

Nitrates are not scale-forming; nor are they a factor in corro-sion problems since they are present only in small quantities.

Nitrite.   Nitrites are most often found only in small quantities if they are present in raw water. They originate primarily from the oxidation of ammonia by bacteria. However, bacteria fur-ther convert the nitrite to nitrate. The presence of nitrite can indicate pollution, but this must be confirmed. In the quantities present in raw water, nitrite is not involved in scale or corrosion.

pH. Most often, the raw-water pH will be in the range of 5.5–8.0. If the pH is above or below this range, contamination is probably present. A low pH can result in corrosion of metals, while a high pH can result in scale formation.

Phosphate. Ground or surface waters seldom contain large amounts of phosphates. If present, phosphate generally indicates fertilizer runoff or pollution. Phosphate from raw water can be the cause of scale problems in open recirculating cooling water systems after the water is concentrated.

Potassium. This cation is only found in small quantities in raw waters except for a few isolated exceptions. Its properties are similar to sodium; its salts are even more soluble than the sodium salts.

Silica. A high silica content, if present, is usually found in ground waters. There, it is not unusual for a silica content of 70–80 ppm SiO_2 to be found. When present in surface waters, silica is often in the colloidal form.

In high-silica regions, silica can cause scale problems in cooling water systems as well as in boilers. Prevention of scale involves maintaining silica in the recommended range.

High pH values aid in keeping silica in solution in both cooling and boiler systems. Silica control is especially important in high-pressure boilers. Silica can vaporize and deposit in the low-pressure areas of turbines as well as superheaters.

Silica is usually the first material to leak in a demineralizer. For this reason, it is used as the control for DI operation. When present as a colloid, silica can leak through a demineralizer.

Sodium. Although present in virtually all raw waters, sodium has little significance in industrial water treatment other than contributing to total dissolved solids. It does not form scale compounds. Where corrosion is concerned, it contributes only by increasing the conductivity.

Sulfate.   This anion is similar to chlorides in that corrosion is aggravated by its presence. Most raw waters contain sulfate. Unlike chlorides, the sulfates are often found in scale, usually as gypsum (calcium sulfate). However, this type of scale can be avoided in most industrial plants through monitoring of sulfate and calcium content.

Total Dissolved Solids.   This is an all-inclusive figure covering all dissolved materials. It is a general indication of corrosiveness of raw water. In plant use, total dissolved solids (TDS) are generally determined by measuring the conductivity of water. This method misses some substances that are not ionized, but it is satisfactory for most plant uses. The conductivity test is a convenient method for detecting a change in raw-water composition. Any major change in a raw-water component is generally indicated by a change in TDS. A measure of TDS by conductivity also serves to indicate cycles of concentration in both cooling and boiler water systems. Total dissolved solids measurement also serves to indicate if carryover takes place in a boiler system.

Total Hardness.   For convenience, calcium and magnesium are usually combined and shown as total hardness ($CaCO_3$), in addition to being shown as Ca and Mg. Combining the two serves to simplify softener calculations, etc.

Part I Pretreatment Performance Problems

For several reasons, raw waters usually require treatment before being added to boiler or cooling water systems. The reasons for pretreatment may be technical, economical, or both.

At the far end of the scale, high-pressure boilers can only operate with essentially pure water. If raw water were used as makeup, the high heat flux would result in failed tubes within a very short period of time. On the other hand, many recirculating cooling water plants operate using raw-water makeup. But to do so, they usually operate at reduced cycles of concentration. At least in larger systems, it is often more economical to use a cold-lime treater to reduce alkalinity, hardness, and possibly silica, iron, and manganese. This not only allows the plant to operate at higher cycles of concentration, but also removes some potentially troublesome ions.

Specifically, pretreatment methods are usually employed to reduce alkalinity, hardness, silica, iron, manganese, turbidity, bacteria, organic matter, etc. As previously mentioned, all ionized solids are removed when makeup to a high-pressure boiler is required.

Pretreatment problems primarily evolve around performance. It must be determined if the equipment is delivering the quality of water specified. In the case where regeneration is involved, as in ion exchange, the question of capacity for a given volume of resin is of concern. The life of the pretreatment materials like ion-exchange resins or reverse osmosis membranes is also important.

Unfortunately, the plant may not always operate with the desired equipment. As a paradigm, many modern plants employ a combination of hot- or cold-lime treatment followed by a zeolite softener. This provides excellent makeup for a medium-pressure boiler; hardness is decreased to virtually zero while alkalinity and silica are reduced. If present, iron and manganese are removed from the raw water. Still, many hot- or cold-lime units exist in older plants without the zeolite softener. These older plants often cannot justify the additional capital expenditure required for modernization. In these cases, the need for proper internal water treatment increases. The results cannot be expected to match those of modern plants. In many cases, increased blowdown or other steps might be required for acceptable results.

1 Ion Exchange

Ion Exchange—Common to Both Cation and Anion Units

Ion exchange is the most widely used method of pretreatment. The major use involves the removal of calcium and magnesium *hardness*. Iron and manganese, if present, will also be removed. For various reasons, however, these two *cations* should be removed before contact with an ion-exchange resin.

Ion-exchange resins consist of an insoluble *polymer* bead to which an exchangeable cation or *anion* is attached. Resins differ from ordinary salts, acids, or bases in that only the cation or anion on the bead is free to enter into a chemical reaction. Accordingly, we have cation and anion resins. The exchange resin has a stronger attraction for the *ions* being removed from the raw water than for the ions present on the regenerated resin bead. In the case of a *zeolite softener* (Fig. 1–1), the attraction for calcium and magnesium is greater than for sodium. Therefore, the resin is capable of removing very dilute amounts

Figure 1-1.   Ion-exchange softener (courtesy of The Permutit Company, Inc., Paramus, N.J.).

of hardness from *raw water*. To reverse this reaction during *regeneration*, very concentrated amounts of sodium chloride (salt) must be used.

In removing cations from the resin during regeneration, either salt or acid is employed as the regenerant. Where acid is used, the effluent from the cation resin bed is a corresponding acid to the anions present, such as hydrochloric, sulfuric, nitric, or carbonic acid.

Correspondingly, anions are removed in the anion bed. Caustic soda is the most widely used regenerant although sodium carbonate (soda ash) is also used.

In the case of a demineralizer, the effluent from the cation unit flows through the anion exchanger in series. Here the *hydroxyl* (OH) group on the resin exchanges with chlorides, sulfates, etc., present in the acid solution. The chlorides and sulfates are bound to the resin beads, and the OH ion combines with the hydrogen ion (H) to form water.

In water treatment, ion-exchange resins are used for a variety of purposes. As mentioned, zeolite softeners are most commonly used in ion exchange to remove hardness. If *alkalinity* removal is desired, a hydrogen cation exchanger can be used. Often, a sodium cation exchanger is operated in parallel with a hydrogen cation exchanger. This is known as a "split-stream" unit.

The vast majority of boiler installations use feedwater pretreated in cation-exchange (zeolite) softeners. The important single function of a salt-regenerated cation exchanger is to remove calcium and magnesium hardness from the raw-water makeup. Cation units are also used on the hydrogen cycle with split-stream units, demineralizers, and weak acid dealkalizers.

Sodium chloride has been mentioned as the regenerant for zeolite softeners. The same salt can be employed to remove bicarbonate alkalinity in an anion unit. In this case, the chloride ion is exchanged instead of sodium.

There are a number of ion-exchange resins available, and they can be used in various combinations to achieve desired results.

Although there are unique properties to cation and anion resins, there are also factors common to both types of ion exchange. These common properties are covered in this chapter.

Probably nowhere in water treatment is recordkeeping of greater importance than in the operation of ion-exchange equipment. Major problems involve the performance and operating life of the ion-exchange resins. Performance covers the quality of treated water plus the capacity per unit volume of the exchange resin. Problems can be caused by many different chemical and equipment defects. Accordingly, any evidence (records, gauge readings, etc.) that aids in pointing to the cause of a problem is well worth the small amount of extra time required of operators or the cost of recording or control equipment.

Basically, records must be kept of the quality and quantity of water treated per regeneration. Gauge readings must also be taken to detect pressure drop across equipment or the resin bed. This information can assist in locating trouble spots.

The definition of quality varies with the type of system employed. In the case of a zeolite system, hardness is the primary factor. On the other hand, virtually pure water is the desired goal of a demineralizer. Quantity covers the amount of water that can be treated by the unit per regeneration and still provide the desired quality. Descriptions of problems that affect ion-exchange operation in general (cation and anion) follow.

Loss of Capacity or Leakage

This problem involves conditions common to both cation and anion resins; conditions unique to cation or anion resins are covered separately.

In investigating ion-exchange performance, it should be kept in mind that problems caused by resins occur gradually while mechanical equipment problems take place rapidly.

Raw-Water Composition Change. Changes in hardness or other ionic components have a very important influence on the capacity of a resin. At times, changes can also exert an influence on the treated-water quality.

To detect a change, it is necessary to make an analysis of the raw water when the ion-exchange unit is first installed. Periodic analyses should follow to determine if any change in the quality of the raw water has taken place.

As would be expected, increases in raw-water hardness decrease the exchange capacity of zeolite softeners. Increases in

conductivity (dissolved solids) of the raw water result in reduced capacity of demineralizers.

Multivalent ions in both cation and anion units are more
readily removed than monovalent ions. Calcium, magnesium,
and *sulfates* are representative multivalent ions, while sodium
and chlorides are common monovalent ions. Accordingly, if
there is a large increase in sodium, the raw water will be more
difficult to treat. If the percentage of multivalent ions increases,
the reverse may be true.

Usually, the composition of ground water will show little
change from year to year. Surface water can exhibit rapid
changes. When a major composition change is indicated, a
duplicate analysis should first be made to verify results. If
correct, the regeneration cycle will have to be changed to reflect
the new composition.

On a daily operating basis, a change in conductivity
measurement (total dissolved solids, or TDS) will usually indicate a change in water quality. Measurements should be made
at least daily on the influent to a softener or other ion-exchange
equipment. If the conductivity increases or decreases, a more
thorough chemical analysis can then be made.

Resin Loss.   As would be expected, if the volume of ion-
exchange resin is reduced, the capacity of the ion exchanger
will be lowered. For this reason, it is important that the resin
bed depth be measured when the resin is first installed. At
subsequent intervals, the bed depths should be recorded.
Strong electrolyte ion-exchange resins should be measured in
the exhausted stage. Weak electrolyte ion-exchange resins
should be regenerated when measured.

If the bed volume decreases, the capacity will gradually
decrease. Quality may or may not decrease, depending on the
type of ion-exchange operation. Cation softening will show litle
decrease in quality until the breakpoint is reached.

Backwash accounts for the greatest loss of resin. Resin that
has suffered no chemical or physical damage can still be lost in

backwash if the rate is set too high. One recommended method of setting the backwash rate is to increase the backwash rate until normal-sized resin beads carry over. This rate is then reduced 3 gpm/ft^2. Or the manufacturer's recommended backwash rate can be used. A small sample line can also be placed on the backwash line. A trap installed on the line will indicate when resin is carried over with the backwash.

Typical backwash rates are 5–7 gpm/ft^2 for a cation unit. Because anion resins are of lower *density* than the cation resins, the required backwash rate is lower. Typical rates are 2–3 gpm/ft^2.

Physical or chemical breakdown of resin can reduce the size of the bead or lessen the density. In either case, the resin will carry over with the backwash even at normal backwash rates.

Backwash rates will also vary with the temperature of the inlet water. When ground water is employed, the rates will change very little. However, surface waters can vary considerably in temperature. In winter, the increased water density requires that the backwash rates be reduced.

Resin can also be lost during the service cycle if screens or piping are broken in the underbed collection system. This can be detected if a sample line and trap are installed in the discharge line. If the break is large enough, the pressure drop from the bottom of the bed and the discharge line will drop. In addition, resin beads may be found in boiler water *sludge* during inspection periods.

All resins will deteriorate over a period of time, but if the loss is excessive, corrective steps should be taken. Regarding corrective measures, the items outlined should be checked. These include backwash rates, condition of resin beads, and possible breaks in the underbed collection system.

Depending on the condition of the remaining beads, additional resin should be added to bring the resin bed back to its original volume, or the entire bed should be replaced.

Meter Operation. When the amount of water being treated per regeneration is either greater or less than normal, the

accuracy of the flow meter should first be determined before
the ion-exchange equipment is investigated.

Channeling.   Channeling is characterized by an uneven resin
surface, often along the sides. Opening a manhole is required
for inspection. When channeling takes place, short cycles and
poor water quality can result. One cause of channeling is the
caking of resin; this is often a result of resin deterioration.
Caking of calcium *carbonate*, dirt, oil, iron *oxide*, etc., can also
occur; clumping from bacterial growth is also a cause of
channeling. Excessive flow rates or pressure drop from excessive *fines* can also contribute. Mechanically, broken dis-
tributors or collectors can cause channeling. Corrective steps
consist of preventing the formation of fines or resin *fouling* and
replacing broken equipment. It can be expected under normal
operation that fines will develop over a long period of time. At
that time, replacement of resin should be made.

Gases.   Carbon dioxide, nitrogen, oxygen, and other gases
can be released in ion-exchange units, resulting in both
channeling and loss of resin during backwashing. When this is
a problem, the gases should be vented manually.

Faulty Regeneration.   Manufacturer's recommendations on
regeneration should be followed. The regenerant solution may
be too weak or too little may be used. The quality of the
regenerant should also be investigated. Salt may contain excessive amounts of insolubles (clay) or calcium or magnesium
salts. Acids and caustics may contain iron oxides. Also refer to
comments under cation or anion regeneration.

Multiport Valve Operation.   Frequently, multiport valves do
not seat. As a result, some of the influent water passes directly
to the discharge without passing through the ion-exchange
equipment. This can be detected by securing an effluent sample after the multiport valve. For example, a cation softener can
be checked for hardness. If hardness is present in the discharge

line from the valve but not directly after the effluent water leaves the bottom of the ion exchanger, leakage is indicated. The valve should be repaired or replaced.

Color Throw. The presence of color in the ion-exchange effluent or *color throw,* can result from several factors. One cause is the storage of resin at high temperatures. Alternate freezing and thawing of the resin is another. Then, too, raw water may contain organic complexes of iron; *bacteria* may also be present. Proper storage of resin is one solution. Where raw-water quality is the cause of color, a pretreatment before the ion-exchange unit, such as cold lime or a carbon filter, is required.

Excessive Pressure Drop/Reduced Flow Rate

Pressure drop normally occurs in any ion-exchange system. Seasonal variations in temperature change the water density, which in turn results in changes in pressure drop. These conditions are to be expected. A commonly encountered maximum pressure drop across the resin bed is 10 psig, plus a distributor pressure drop of 2–4 psig (Owens, 1985).

Greater than normal pressure drops over an ion-exchange unit present a problem. The end result is a decrease in flow rate. If the pressure drop is caused by resin fouling, channeling may occur. If this takes place, the pressure drop may then decrease and the quality of the treated water also decreases.

Increased pressure drop can result from high flow rates, large amounts of fines, plugged or broken inlet distributor systems or underbed collection systems, or resin fouling. *Pressure gauges* can assist in determining the cause of excessive pressure drop. Gauges should be placed at least at the inlet and outlet to an ion-exchange unit. Gauges are also recommended at the top and bottom of the resin bed. These four gauges will indicate if the pressure drops are occurring through the resin bed or through the distributing or collection sections.

Corrective steps consist of replacing defective equipment, replacing or cleaning the resin bed, and following the manufacturer's recommended service flow rate. Usual service flow rates are 6–8 gpm/ft^2.

Ion Exchange—Cation

Loss of Capacity or Leakage

Physical and Chemical Deterioration. Cation resins tend to deteriorate from physical causes, while anion resins usually fail from chemical attack. Specifications for new resins should be obtained from the manufacturer. This will serve as a basis for future evaluation of the resin. In addition, photographs of the new resin should be obtained. Photographs should be taken on a yearly basis to determine the amount of change occurring in the resin bed. This service can be provided by many water treatment firms. Sampling is important to obtain useful information. One excellent method is to use a *grain thief,* or tubular device that enables samples to be taken evenly throughout the entire bed depth.

Fines will normally form over a period of time. Attrition develops not only from the rubbing of resins against each other but also from the swelling and shrinking of resin during exhaustion and regeneration. If the fines form at a rapid rate, the operator should be concerned and investigate the cause. Alternate freezing and thawing of resins during idle periods can result in resin deterioration. Necessary steps to prevent freezing should be taken.

Excessive pressure drop, often the result of high flow rates, can also result in the physical breakdown of resins. The specified flow rates should be followed if at all possible.

Free chlorine or oxygen can cause chemical deterioration of cation resins. Cation resins on the hydrogen cycle are more readily attacked by chlorine than neutral solutions. In either case, chemical attack is centered on decrosslinking. Crosslinking of resins provides physical strength. Total capacity of the cation bed does not normally deteriorate during the life of the bed.

If the moisture content exceeds the original by 15% or more, the resin should be replaced.

Catalyzed sodium sulfite can be used to reduce chlorine before the water enters the resin bed. If sodium sulfite is not used, the free chlorine should be held below 0.3 ppm. Decrosslinkage is characterized by increased moisture content. This moisture content determination can be made by a qualified water treatment laboratory. The resin also becomes soft and jellylike to the touch.

The most widely used cation resins are resistant to high temperature up to about 280°F. Accordingly, they can be used after hot-lime softeners. These resins are even stable at high temperatures when oxygen is present in the water. But the resins do not resist the combination of high temperature, oxygen, and dissolved and suspended iron.

Condensate frequently contains oxygen and traces of soluble or suspended iron oxides. If this water is blended with a hot-lime effluent, the life of the resin can be shortened. When the combination of high temperature, oxygen, and iron is present in the influent to a cation resin exchanger, steps should be taken either to remove the oxygen or, by *filtration*, to remove the iron.

Fouling of the Resin.   Cation resins usually *foul* from inorganic deposits, while organic contaminants are responsible for most anion resin fouling. The most common and troublesome cause of cation resin fouling is the presence of iron in the influent. If no oxygen or other oxidizing agent is present, the iron can be exchanged in the same manner as calcium or magnesium. In most cases, however, the iron precipitates on the resin. This usually takes place only in brine-regenerated plants. Where acid is used, the iron is effectively kept in solution (Duolite Staff, 1982, p. 42).

Even if oxygen is not present in the incoming water, it is present to a reduced degree in the regenerating solution and often in the backwash water. This can result in the oxidation of iron to the insoluble ferric hydroxide within the beads. Copper and manganese can foul resin in a similar manner.

The best solution is prevention. It should be kept in mind that an ion exchanger is not a filter. Metal salts of iron, manganese, or copper should be removed before entering the ion exchanger. The feeding of polyphosphates has been reported to cause a less dense precipitate in iron-bearing water, resulting in better removal during backwash (Nachod and Schubert, 1956). For iron control, the usual recommended dosage is 2 ppm polyphosphate (PO_4) per 1 part of iron plus a base dosage of 2 ppm polyphosphate.

Occasionally, the aluminum ion will deposit in cation beds. It is usually found only in municipal water systems, where it is used as a coagulant. Since it is only present in small amounts, it reduces the capacity of a resin gradually. The aluminum ion, being trivalent, is strongly bonded to the resin and is difficult to regenerate with salt.

Metal hydroxides can often be removed by cleaning with hydrochloric acid (5–15%), phosphonates, or sodium bisulfite. One recommended procedure specifies the use of 10% HEDP phosphonate in water solution.

Recommended dosage is 1 gal of product per ft^3 of resin. Soak for 1 h. One method of cleaning the resin bed is to remove the top manhole cover. Then, backwash until clear solution appears. Drain down to about 1 in. above the bed. Add the required amount of resin cleaner. "Air lance," soak, and repeat backwash until clear solution results.

If the bed contains a large amount of fines, cleaning will probably not improve performance a great deal. Replacement with new resin should be considered.

Fouling from calcium *carbonate* or magnesium hydroxide is quite common in units that employ the hot-lime/zeolite combination pretreatment. Corrective measures lie in improving the hot-lime operation to prevent carryover or in cleaning the filters following the lime treater. Surface waters often contain suspended matter and organic acids. These products can foul resins, and they should be filtered before they enter a cation unit. Fouling can also take place when a poor grade of salt or acid is used for regeneration. High calcium and magnesium contaminants interfere with regeneration. The salt should contain a high sodium chloride content and be free from dirt. Acid should be free from contaminants.

If sulfuric acid is employed, care must be taken that calcium *sulfate* does not precipitate within the beads. If this is a problem, stepwise regeneration should be used. Regeneration starts at 2% and increases to 4 or 5% acid. Hydrochloric acid avoids the problem but is more expensive than sulfuric acid.

Normally, a large percentage of deposits on resin beds can be removed by backwashing, but the backwash rate must operate in the proper range. If the flow rate is greater than recommended, resin is washed out, along with any deposits present on the resin bed. On the other hand, if the flow rate is too low, deposits on the bed are not removed.

Resin manufacturers usually specify a flow-rate range for backwash in order to expand the bed sufficiently and remove deposits. Traps are frequently installed on sample backwash lines to detect overflow of resin. Backwash rates vary at different temperatures since water is denser at low temperatures. This is not a problem using ground water since the temperature usually stays fairly constant; it can present difficulties when surface waters are used. Resin manufacturers usually specify backwash flow rates at different water temperatures.

Bacterial growth can also foul cation beds. To sterilize the resin bed, formaldehyde is often used. One volume of 40% formaldehyde should be diluted with 40 volumes of water (Duolite Staff, 1982, p. 50). Normally, the formaldehyde can be injected using the regeneration batch preparation tank. Other proprietary *biocides* have been used by water treatment companies. In the selection, care must be taken that toxic products are not used when the treated water is required to be potable-grade. Formaldehyde is a toxic chemical.

Excessive Rinse. If a cation resin bed is not rinsed to the desired level, high chlorides or sulfates result, depending on the operation. The result is increased total dissolved solids (TDS) in the boiler feedwater. If a higher than normal rinse is required, it is often an indication of resin deterioration. The resin should be checked by a water treatment company or resin manufacturer and replaced if necessary.

In zeolite softener operation, an increase in conductivity (TDS) between the influent and effluent waters indicates the presence of chloride in the treated water. Chlorides can also be tested for directly in the influent and effluent waters.

Faulty Regeneration. If a cation resin is not regenerated properly, optimum capacity or quality of treated water cannot be achieved.

In water softening, the most widely used ion-exchange operation, the capacity is determined primarily by the regeneration level. This is the pounds of salt used per regeneration per ft^3 of cation resin. The usual efficient level of salt is 6 lb of salt per ft^3 of resin.

While the maximum concentration of salt is about 26%, this concentration can dehydrate the resin, resulting in cracking. If a very dilute solution is used, enough contact between the salt and the resin is not made, resulting in loss of capacity. The typical salt concentration used is 10%. This can readily be checked with a salometer. Typical regenerating flow rate is 1 gpm/ft^3 of resin for 20 min.

Usual concentrations for other cation regenerants are hydrochloric acid, 4–10%, and sulfuric acid, 2% plus. Flow rates for hydrochloric acid are 0.5–1 gpm/ft^3 and for sulfuric acid 0.5–1.5 gpm/ft^3. In all cases, the manufacturer's recommendation should be followed.

A record of the quantity of regenerant used vs. number of regenerations will determine if a change has occurred in the amount of regenerant used. The quality of salt or acid should also be checked. This can be determined by observation or chemical analyses. If too high a concentration of sulfuric acid is used, calcium sulfate forms on the beads. During exhaustion, calcium leakage appears initially and gradually decreases. Refer to the section on Fouling of Resins, page 28.

When brine is used for regeneration, high chlorides may appear in the effluent. Refer to the foregoing section on Excessive Rinse.

With weak acid cation resins, the alkalinity in the effluent is sometimes above specifications. Most often, this is caused by a low level of regeneration. The amount of regenerant should be increased.

Service Run Conditions.   Temperature and flow rates are two factors that can influence the quality and quantity of cation treated water. High flow rates have little influence on zeolite softener operation. Normal flow is in the 5–10 gpm/ft^2 range. Weak acid cation resins are more sensitive to higher than normal flow rates. Manufacturer's recommended flow rates should be followed.

Temperatures in normal ranges have little effect on strong cation operation. Capacity for weak acid cation resins, however, does change with temperature. At 75°F, the capacity for one cation resin increased 20% over operation at 55°F.

Rapid changes in temperature should be avoided. For example, the resin might exhaust under hot conditions and be regenerated cold.

Little corrective action can be taken for different flow rates or temperatures except to adjust the length of the service cycle.

Ion Exchange—Anion

Loss of Capacity or Leakage

In most water treatment applications, an anion exchanger is combined in series with a cation unit to form a demineralizer. Anion resins also form mixed bed units with cation resins. On smaller industrial installations, cation and anion units are placed in series to soften and dealkalize raw water. For both units, salt is used as the regenerant.

Anion resins have many of the same problems as cation units. These are covered earlier in this chapter. Anion resins also have some unique problems. The many combinations of cation and anion exchangers cannot be covered, but many of the common problems are listed.

Physical and Chemical Deterioration.   Anion resins usually experience a shorter life than cation resins. An average life can vary from 0.5–2 million gal/ft^3 of resin. While cation resins usually fail from physical attrition, the anion resins are more prone to fail from chemical attack. Exchange capacity of the resin gradually drops during its operating life.

As mentioned, anion resins are usually used in the demineralization process in series with cation units. Changes in the performance of the resins can often be detected by changes in *pH*. For example, when strong acid cation and strong based anion resins are employed, the effluent is about 8–9. If the pH rises, the cation bed should be investigated. With a drop in pH, the anion bed should be checked.

Attack on the resin is usually the result of oxidation, i.e., chlorine or oxygen. Consideration can be given to the addition of sodium sulfite if the life of the resin is shorter than expected.

Since anion resins have a lower density than cation resins, greater care must be taken during backwashing to prevent loss of resin. As with cation resins, specifications for new anion resins should be secured from the manufacturer. Also, photographs of the resin should be secured yearly to detect any physical changes.

If fines develop, loss of resin will occur at fast or even normal backflow rates. These fines have to be replaced to restore the bed to normal capacity. Anion resins differ from cation resins in that that they cannot withstand elevated temperatures. Strong base anion resins, for example, should not operate at a temperature higher than 95°F. Within their operating range, the weak base anion resin exchangers perform more efficiently at high than low temperatures.

As mentioned, anion resins have a shorter life than cation resins. Strong base anion resins are considered spent when their ability to remove silica or bicarbonate decreases by 50% (Leferre, 1982). Whenever silica leakage develops, it should not be assumed that the anion bed fails; the cation bed may not be operating properly (Peters, 1982), which adversely affects the performance of the anion bed.

Fouling of Resins.   Anion resins are more susceptible than cation beds to organic fouling. Humic acid from surface water is the most common source. Organic fouling can sometimes be detected by a drop in the pH in the effluent. Final leveling off will be in the range of 4–6 pH. Corrective measures include carbon or sand filters, prechlorination, or the use of macroreticular resins, which resist organic fouling.

Degradation products from cation beds are another cause of anion bed fouling. A 10% sodium chloride–1% caustic solution has been used to clean organics from anion beds.

Although organic chemicals are the usual source of fouling in anion beds, some cases of inorganic fouling also occur. Iron salts, for example, can form complexes and pass through a cation unit and exchange in the anion beds. There they can oxidize in the beds if the treated or regenerating water contains oxygen. Catalyzed sodium sulfite can be used to remove the oxygen and prevent the formation of ferric hydroxide.

Caustic regenerant solution can cause deposits on anion beds in another manner. If hard water is used, magnesium hydroxide can deposit on the resin. Impure caustic is another source of iron that can foul the anion resin. Corrective measures are to replace hard water with soft water and use an iron-free grade of caustic for regeneration.

Silica can foul an anion bed by polymerizing within the bed. It can be removed by treating with warm 115°F caustic solution. Occasionally, oil will contaminate a resin bed. Warm caustic soda plus a detergent can be used to correct the problem.

Excessive Rinse.   This operating condition can result in some loss of capacity. As with cation resin operation, long rinse time is usually an indication of resin deterioration. If the bed is not rinsed sufficiently, high caustic alkalinity will result in the effluent.

Channeling of the resin bed can result in long rinse periods. This, in turn, can cause shorter runs.

Faulty Regeneration.   Certain specific conditions apply to anion, as well as cation regeneration. Usually, the anion problems involve silica, which is not highly ionized. Silica is more readily removed if the caustic regenerating solution is heated to 95–125°F.

If silica is not completely removed, there is danger that it will concentrate and polymerize within the strong base beads (Davies, 1984). During the run, it can then appear as leakage.

If the weak base anion resin is regenerated in series with the strong base resin, problems can sometimes be encountered. The initial spent caustic solution from the strong base resin can contain high amounts of silica, which can then deposit in the weak base resin as the pH is lowered. The solution is to discard the initial 15–30% of the spent caustic solution from the strong base resin bed.

Finally, to avoid precipitation of hydroxide or carbonates, dilution and rinse water used in the regeneration cycle should not contain hardness. The caustic used for regeneration should be rayon-grade.

Service Run Conditions.   The main factors to be considered during the anion service run are temperature of the incoming water, kinetic loading, and flow rate.

In the usual temperature range of 60–80°F, the capacity or quality of anion-treated water is not greatly affected. However, in winter months, when the temperature of surface waters may drop lower than 60°F, the weakly basic anion resin may be adversely affected.

When low silica levels are specified, a temperature increase in the service water can result in increased leakage (Castagna and Miller, 1981). One reference mentions a 50% increase when the temperature was raised to 85°F from 70°F. In all cases, temperatures should operate in the range recommended by the manufacturer if possible.

Regarding kinetic loadings, the anion resins are affected to a greater degree than strong acid cation resins.

Strong base anion resins do not exhibit sharp drops in capacity or quality if flow rates increase. The opposite is true for the weak base exchanger resins, which react slowly. Silica removal by strong base resins is the exception. If maximum silica reduction is desired, high flow rates should be avoided. Check manufacturer's recommendations. Low flow rates can also cause channeling in anion exchangers. Flow rates should not fall below 2 gpm/ft^2.

When anion resins are stored for over a month, the resin should be left in the exhausted stage.

Usually, the operating conditions contributing to service run problems cannot be changed. Nevertheless, to maintain the desired effluent quality, it may be necessary to shorten the time between regenerations.

2 Hot- and Cold-Lime/Soda Process

Hot-Lime/Soda Process

High Hardness, Alkalinity, or Silica Effluent

The hot-lime/soda process (Fig. 2-1) is an effective method of removing alkalinity as well as reducing hardness and silica.

Quick lime or hydrated lime reacts with calcium bicarbonate in the following manner:

$$CaO \text{ (quick lime)} + H_2O \rightarrow Ca(OH)_2 \text{ (hydrated lime)}$$
$$Ca(OH)_2 + Ca(HCO_3)_2 \rightarrow 2\ CaCO_3 \downarrow + 2H_2O$$

Magnesium bicarbonate removal requires additional lime.

$$Ca(OH)_2 + Mg\ (HCO_3)_2 \rightarrow CaCO_3 \downarrow + MgCO_3 + 2H_2O$$
$$MgCO_3 + Ca(OH)_2 \rightarrow Mg(OH)_2 \downarrow + CaCO_3 \downarrow$$

If noncarbonate hardness is present, the addition of soda ash is required; for magnesium removal, lime is also needed.

Figure 2-1.   Hot-lime/soda softener (courtesy of The Permutit Company, Inc., Paramus, N.J.).

$$CaCl_2 + Na_2CO_3 \rightarrow CaCO_3 \downarrow + 2\ NaCl$$
$$MgCl_2 + Ca(OH)_2 \rightarrow CaCl_2 + Mg(OH)_2$$

The $CaCl_2$ formed in this reaction is removed by the addition of more soda ash. Since the lime/soda process reduces hardness only to the range of 6–20 ppm $CaCO_3$ vs. <1 ppm for ion exchange, it is often combined in series with a sodium cation exchanger. The ion-exchange softener follows the hot-lime softener. This combination has been made possible by the development of high-temperature-resistant ion-exchange resins. When combined with ion exchange, the sodium carbonate (soda ash) is usually eliminated.

The use of salt rather than soda ash in regenerating the cation-exchange softener is a less expensive method for removing noncarbonate hardness. The use of an ion-exchange softener also reduces the total hardness level to below 1 ppm $CaCO_3$. Hot-lime/soda softeners without ion exchange are usually restricted to older plants. A few hot-lime plants, followed by external hot-phosphate treaters, are also in existence.

Hot-lime or hot-lime/soda softeners are usually restricted to larger industrial plants using highly alkaline, hard raw-water influent. Also, these waters often contain high levels of silica, which the hot-lime treatment can remove.

The limitations of the hot-lime or hot-lime/soda process are several. Equipment employed is relatively large, which results in high capital expenditure. Sludge handling represents a major limitation of the process. *Scale* formation in the unit and the feeding equipment is also a major problem. The hot-lime process lends itself best to steady loads with the use of a consistent-quality raw water. Rapid changes in load or raw-water quality can result in sludge *carryover* or poor-quality effluent. Accordingly, close supervision of the process is required.

In operation, raw water is sprayed into the upper chamber and mixed with steam. Oxygen and carbon dioxide are released and vented to the atmosphere. Lime and soda ash (if used) are also added to the upper chamber. The chemical reaction is almost instantaneous except for silica removal. The bottom portion of the upper chamber is cone-shaped.

Reacted water is withdrawn at the bottom of the cone and passed through the *downcomer* vertical pipe to the bottom of the softener. The bottom of the softener is also cone-shaped. Here a sludge bed is allowed to develop. The reacted water from the upper chamber discharges to the bottom of the unit and reverses its flow as it passes through the sludge bed. The sludge bed allows enough contact time for the reacting chemicals to *flocculate*. Silica is adsorbed on the sludge bed. Clarified water above the sludge bed is withdrawn near the top of the hot-lime treater and pumped to the filter. As required, sludge is withdrawn from the bottom of the softener. In addition to its main functions, the hot-lime/soda process is an effective method of removing iron and manganese, as well as turbidity.

Low Temperature.   The most important factor in the successful operation of a hot-lime/soda treater is temperature. Usually the temperature of the effluent is slightly above 220°F., depending on the steam pressure. If the temperature is low, the steam regulator should be checked, as well as the flow rate of the water being treated. Under normal conditions, the water temperature should be within 1–3°F of the steam temperature. Failure to achieve this temperature may indicate spray problems.

High or Erratic Flow Rate.   If the flow rate through the unit is higher than designed or is erratic, it may be difficult to maintain sludge bed levels. High hardness from sludge carryover will result; other components normally found in the sludge will also be present in the effluent. Silica requires more reaction time than calcium or magnesium. Accordingly, at high flow rates, high silica readings can be expected. In an efficiently operated unit, silica can be expected to drop to 1–3 ppm.

Where high flow rates prevail, it should be determined if the plant can operate on a more even basis. Another possible alternative is to add a storage tank to level out demand.

Chemically, a *coagulant* or synthetic polymer might be added to produce a denser bed. If a polymer is being used when poor results are obtained, a change to a different type of polymer should be considered; or the dosage should be adjusted.

If a unit is vastly underdesigned for the required flow rates, poor results can be expected, regardless of other changes made.

Channeling.   Incrustation of calcium carbonate and/or magnesium hydroxide can cause channeling. It is also possible that too dense a sludge bed was produced through overdosage of an inorganic coagulant or synthetic polymer. Even in normal operation, some buildup of deposit will result over a period of years. In all cases, the deposits should be removed mechanically during shutdown periods. If buildup is excessive, a change in coagulant or synthetic polymer should be considered.

High Sludge Bed.   Under normal operation, the sludge bed should be 6–8 ft in height. If the bed is carried higher, the chance of carryover of the sludge is increased.

The bed should be maintained in the normal recommended operating range.

Vent Condenser Leakage.   Some hot-lime/soda softeners incorporate an external vent condenser to preheat the raw water and remove unwanted gases. If leaks develop in the tubing, it is possible for raw water to bypass to the effluent water through the vent condenser drain. The possibility of this occurring depends on the design of the particular unit. To check for leaks, a sample line should be installed in the vent condenser drain line. High total dissolved solids in the vent condenser drain indicate leakage. Necessary replacement of tubes should be made.

Poor Flocculation.   The quality of the *flocculation* depends on temperature, chemical control, coagulant or polymer selection, plus mechanical factors, which have been discussed. Proper chemical dosage should result in effluent containing a hardness of 6–20 ppm $CaCO_3$. Lime dosage is correct if the hydroxide alkalinity is 0–10 ppm ($CaCO_3$). If soda ash is employed, the dosage is correct if the total alkalinity M is 30–40 ppm in excess of the total hardness as $CaCO_3$ (McCoy, 1981 and Hamer, Jackson, and Thurston, 1961).

Polymers and/or sodium aluminate are usually not required in a hot-lime/soda treater. If conditions require a flocculant, bench tests should be made to determine the choice of product and required dosage.

The clarity of the effluent is influenced by:

1. The upward velocity of the treated water in the area where water and solids separate. This is influenced primarily by the flow rate on the softener.

2. Temperature. Separation of water and solids is more readily made at higher temperatures.

3. Height of sludge bed.

4. Density and size of solids. These are influenced by raw-water quality, polymer selection, and dosage, etc.

Flow-control fluctuations can upset the sludge beds, but the sludge recirculation line tends to dampen this upset. It should be determined that this line is open.

Erratic or Incorrect Chemical Feed.   Although effluent analyzers have been used to control chemical feed, their use is the exception.

The large majority of hot-lime treaters feed chemicals in proportion to flow. A mechanical or electrical problem in feeding cannot be corrected by water chemistry, but the appropriate action should be taken. Plugged lines from the lime-soda

slurry tank will also interfere with chemical feed. Sodium ligno-
sulfonate has been used successfully to prevent precipitation in
slurry tanks and lines; about 1 lb per 300 lb of lime is recommended. Dosage can vary, depending on results. Various
polymers have also been used to control scaling.

Recommended feed dosages are as follows: Hydrated lime
(93% $Ca(OH)_2$); number of pounds per 1000 gal of water treated
= (Alk + Mg)/150, where Alk equals alkalinity, ppm $CaCO_3$,
and Mg equals magnesium hardness as ppm $CaCO_3$. Requirements for soda ash, 98% Na_2CO_3 are as pounds per 1000 gal of
water treated = (H − Alk)/110 plus required excess; H equals
total hardness as ppm $CaCO_3$.

Silica is removed by reaction or adsorption with precipitated
magnesium. The required magnesium will normally be present
in the raw water. In those cases in which the magnesium is
low, additional magnesium will have to be added either as
hydrated dolomitic lime, $Ca(OH)_2MgO$, or magnesium oxide.
Dolomitic lime is the least expensive source of magnesium.
Both the dolomitic lime and magnesium oxide may be added
from the standard lime and soda ash wet chemical feeders. If all
normal operating factors are present, the silica in the effluent
should be reduced to 1–3 ppm.

Changes in Raw-Water Quality. As would be expected,
changes in raw-water quality can have an adverse effect on the
quality of effluent water. A complete analysis of raw water
should be made at least once per year. Changes in chemical
feed should be made to compensate for raw-water changes.

The problems caused by minor changes in ground-water
quality are of little concern when compared with those units
that employ surface water as *makeup*. Some small rivers seem to
change quality on a daily basis. When plant personnel operate
under such conditions, a less than perfect effluent quality is to
be expected. Changes in raw-water quality can be indicated in
daily operation by checking the conductivity (total dissolved
solids, or TDS) of the water. If there is a drastic change, a more
complete analysis can then be made.

Most operators have found that better results are obtained when gradual changes in chemical dosages are made to adjust to rapidly changing raw-water quality.

Defective Water Sprays. The more intimate the contact between the raw water and the steam, the more rapidly the temperature will rise. For this reason, it is desirable that the sprays operate efficiently.

When the lime/soda treater is off-line, sprays should be examined for plugging or weak sprays. Required corrections should be made.

Damaged Lagging. When lagging is damaged, different temperatures exist in the hot-lime/soda treater. This can cause irregular flow in the sludge bed. Defective lagging should be repaired.

High Aluminum

When sodium aluminate is used as a coagulant, it appears to be effective only when the magnesium content is 30 ppm or greater (Hamer, Jackson, and Thurston, 1961, p. 9). At low magnesium content, the aluminum tends to remain in solution.

When aluminum is found in the effluent, its use should be discontinued. A change to an organic flocculant can be considered as a replacement.

High Iron or Manganese

The hot-lime/soda treater is an excellent vehicle for removing iron and manganese. If iron or manganese are found in the effluent, it usually means that excessive carryover of sludge has taken place. Refer to High Hardness, Alkalinity, or Silica Effluent, page 39.

It is also possible for the raw water to bypass the hot-lime/soda treater. Refer to the section headed Vent Condenser Leakage, page 35.

Cold-Lime/Soda Process

Although the basic chemical reactions are the same as the hot-lime/soda process, the equipment is different and, of course, the temperature is lower. As a result, reactions do not go as far to completion as in the hot-lime/soda treaters. Flocculation is also more difficult.

High Hardness, Alkalinity, or Silica Effluent

Cold-lime/soda softeners (Fig. 2-2) are normally used as pretreatment for cooling water systems, process, and boiler feedwater. Usually, the cold-lime process is only used for boiler

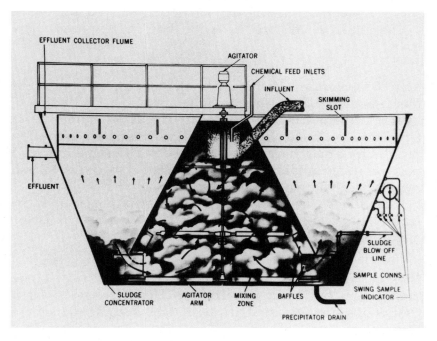

EFFLUENT COLLECTOR FLUME

AGITATOR

CHEMICAL FEED INLETS

INFLUENT

SKIMMING SLOT

EFFLUENT

SLUDGE BLOW OFF LINE

SAMPLE CONNS.

SWING SAMPLE INDICATOR

SLUDGE CONCENTRATOR

AGITATOR ARM

MIXING ZONE

BAFFLES

PRECIPITATOR DRAIN

Figure 2-2.   Cold-lime/soda softener (courtesy of The Permutit Company, Inc., Paramus, N.J.).

pretreatment when it is also being used for cooling or process pretreatment. The primary function of a cold-lime/soda process is to lower hardness and alkalinity. It will, in addition, reduce silica slightly; iron and manganese will be removed. To remove manganese, enough lime must be added to raise the pH above 10.0. Turbidity present in surface waters is usually removed, and bacterial count is lowered substantially.

If lime alone is used, only the calcium in the bicarbonate form is removed. When equal or excess alkalinity is present in the raw water, the expected effluent calcium will be about 35 ppm (as $CaCO_3$). A 10% reduction of magnesium (as $CaCO_3$) also will take place.

A small amount of silica will also be removed, depending on the reaction time and amount of $Mg(OH)_2$ present in the sludge. Potential problems that follow are similar in some cases to the hot-lime/soda process, with different emphasis, however.

Low Temperature.   If the raw water originates from wells, there should be little variation in temperature. Depending on location, there can be seasonal changes in surface-water temperatures. If aluminum or iron coagulants are employed, the dosages may have to be increased. The same applies if polymer flocculants are used. The increase in hardness, alkalinity, or silica will result from carryover.

It should be kept in mind that temperature is only one variable in cold-lime/soda operation; other seasonal factors might improve flocculation.

High or Erratic Flow Rates.   As expected, flow rates higher than design can result in carryover. High flow rates can be continuous or intermittent. If intermittent, a storage tank might be employed to even out the flow at a lower, even rate.

If flow rates cannot be lowered, then jar tests should be performed to determine if a denser sludge can be obtained through the use of a different coagulant or flocculant.

Channeling.   Incrustation of calcium carbonate or magnesium hydroxide in the cold-lime/soda treater can result in carryover. The location of deposits varies with the particular equipment employed and/or the water quality. It is also possible for channeling to result if some part of the equipment fails. Build-up of scale or equipment failure can be determined only by examination during major shutdown periods.

Channeling can also result if too dense a sludge bed is produced from incorrect selection or dosage of a coagulant or flocculant.

High Sludge Beds.   Carryover of sludge can result if the bed is carried at a higher level than design. Maintaining the sludge bed at the correct level is necessary for successful operation.

Poor Flocculation.   Optimum flocculation depends primarily on the proper selection of flocculants. The selection should be made on the basis of jar test results. Depending on plant results, adjustments can be made in dosages. Inferior flocculation can also result from organic contaminating agents. This can originate from humic or fulvic acids present in surface waters.

Locating the source of organic contamination can often prove difficult. In one case history, the elusive contaminating product in a cold-lime treater proved to be the blowdown from a boiler (Schroeder, 1982). No source of contamination should be overlooked.

In many cases, prechlorination can remove interfering organics. The possible benefits of chlorine can be determined in bench tests before it is used on plant scale.

Erratic Chemical Feed.   The remarks made for hot-lime/soda treater, pages 31–38, also apply to the slurry tank used in cold-lime/soda applications.

Raw-Water Quality.   Rapid changes in raw-water quality can make control of chemical feed difficult. Ordinarily, ground water does not present this problem, but some surface waters can change frequently.

In addition to rapid changes, some raw waters have organics that contain complex calcium or magnesium. A few plants in this country use sewage water effluent as makeup to a cold-lime/soda treater. Using this quality of raw water, it is not uncommon for the calcium hardness in the effluent to be higher than the influent water. The only reason for using a cold-lime/soda treater in this case is the removal of phosphate as calcium phosphate. The phosphates must be removed since they present serious scale problems in recirculating cooling water systems. In this case, the operator can only accept the fact that the *makeup water* cannot be softened.

Where the raw-water quality changes rapidly, alterations in chemical feed must be made. However, changes should be made slowly to avoid overcompensation.

3 Reverse Osmosis Operations

Although the principle of reverse osmosis (RO) (Fig. 3-1) has been known for many years, only in recent years has it been utilized to any great degree. In nature, osmosis involves a process in which water passes through a semipermeable membrane from a low to a higher concentration. This process can be reversed by applying pressure on the raw feedwater side of a selective membrane. The membrane divides the raw water into concentrated and dilute portions. A simplified way of viewing an RO unit is to consider it as a conventional pressure filter. Unlike an ordinary filter, an RO membrane rejects soluble ions as well as particulate matter. Another difference is that an ordinary pressure filter operates until buildup of particulate matter necessitates downtime while the filter is backwashed. Reverse osmosis units are intended for continuous operation; only part of the water flow passes through the membrane. The remaining concentrate is discharged to waste. This part of the operation might be compared to the *continuous blowdown* of a boiler. By design, turbulent cross flow at the membrane surface

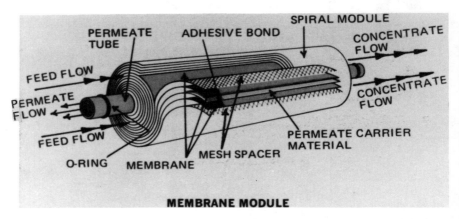

Figure 3-1.   Reverse osmosis: membrane module (courtesy of Osmonics, Minnetonka, Minn.).

assists in preventing buildup of ions and suspended matter. If a buildup of ions were allowed to accumulate at the membrane surface, a point would be reached at which insoluble salts such as calcium carbonate and calcium sulfate would form a scale.

While reverse osmosis appears to be a simple operation, many factors in its operation have to be considered, among them: feedwater composition, pretreatment requirements, type of system design, type of membrane, types of possible fouling, cleaning requirements, and posttreatment specifications.

In recent years, new technical developments have been rapid; they can be expected to continue. Improvements will tend to increase membrane resistance to chemical and physical deterioration. Since a main operating cost involves high-pressure pumps, steps are being taken to reduce pumping pressure requirements.

There are two design types of RO units: spiral-wound and hollow-fiber. The spiral-wound design has a membrane wrapped around a central collection tube. Flow channels are provided by spacers. The spacers provide turbulence, which in turn assists in preventing buildup of ions and particulate matter on the membrane surface.

The hollow-fiber RO unit consists of fibers installed in a sealed module. Water is pumped into the modules with little turbulence, resulting in a greater tendency to foul from particulate matter. Accordingly, pretreatment requirements are higher for the hollow-fiber design. On the other hand, hollow-fiber modules provide the highest ratio of surface area to volume.

Poor Performance

Temperature. When cellulose acetate membrane is used, higher temperatures increase flux or production rate of the permeate. However, membrane life decreases so that an optimum range of 75–85°F. is usually maintained. The operating temperature should not be allowed to exceed whatever limit is specified by the manufacturer (Kremen, 1970).

Low temperatures can cause a substantial reduction in the rate of water production. Unfortunately, on a daily basis, the operator has little control over influent water temperatures. In the long view, it might be possible to increase the raw-water temperature. For example, an elevated temperature waste water could be used through a shell and tube heat exchanger to preheat the incoming water.

The amount of change to be expected in water production for a given temperature should be determined from the manufacturer.

Fluctuating Demand. Reverse osmosis performs best under continuous operating conditions. In this respect, it differs from ion exchange, which is available on demand. Reverse osmosis units should operate for at least 30 min at a time. If loads fluctuate, provisions should be made for a storage tank.

Scale Formation. As ions concentrate at the RO membrane surface, a number of salts can reach their solubility limits. The most widely occurring are calcium carbonate and calcium sulfate. Also included are silica and calcium silicate compounds.

Scale is one type of possible membrane deposit. Other types of deposits include suspended matter, colloids, and microbiological growth.

Control of scale problems begins with a water analysis of the feedwater. This will indicate the tendency of certain ions to participate in the formation of scale on the membrane surface. One control is to reduce the recovery rate of the RO unit. This, of course, will reduce the efficiency of the operation. The most common scaling salt is calcium carbonate, and this can be controlled by reducing the pH of the feedwater. This converts carbonates to bicarbonates, which are more soluble. A common control range is pH 5.5–6.0 using sulfuric acid. When using acid, it is necessary to stay within the recommended pH operating limits specified for the membrane used. Excess acid can damage the membrane and cause corrosion problems on metallic parts.

Certain chemicals are capable of controlling some deposits by dispersing the insoluble salts or interfering with their crystalline growth. These chemicals are effective in very small dosages. They are not effective on all types of scale, particularly silica.

Chief among these products are *polyphosphates*, phosphonates, *polyacrylates*, and many proprietary chemicals. A common .concentration of polyphosphate is 4–6 ppm. Iron and manganese, too, can precipitate on the membrane. Polyphosphate can also be used to control iron and manganese if the levels of the metal ions are 1.5 ppm or below.

Silica, when combined with calcium, can be prevented from forming scale by controlling the calcium concentration. There are no in-line methods of controlling pure silica scale if the solubility concentration is exceeded.

When a scale problem exists, the deposit should be analyzed to determine its identity accurately. To detect scaling (or other type of deposits) during operation, the flow rates of the permeate and concentrate should be kept as constant as possible. By so doing, it is possible to detect plugging, which results in an increase in differential pressure.

Cleaning solutions should be used when the feed-to-concentrate differential pressure drop increases by 10–20%. Pure silica scale cannot be removed by cleaning solutions. On a routine basis, RO systems should be cleaned at least twice a year (Kaup, 1973). Operation results may indicate a different schedule.

The manufacturer should be contacted so that no damage results from improper pH, temperature, contact time, cleaner concentration, etc. Selection of the proper cleaning agent will also be determined by the type of deposit.

If scaling cannot be prevented by regulating salt concentration at the membrane, chemical additives, or cleaning, then some type of pretreatment must be considered. The most common type of pretreatment is the addition of a softener to remove calcium and magnesium salts.

Suspended Solids. Particulate and colloidal matter in raw water can adversely affect membrane performance. The spiral-wound configuration is better suited for handling particulates and colloids than hollow-fiber modules. Filters are commonly used as pretreatment for RO systems. Diatomaceous earth should not be used in pretreatment filters. Any carryover to the membranes can seriously interfere with their operation.

Raw-Water Quality. References have been made to ions in feedwater that can cause RO system problems. It is important that a chemical analysis be made of this water. Of greatest importance are the following ions: calcium, magnesium, sulfate, carbonate, barium, strontium, iron, manganese, and aluminum. All can be involved in scale formation. Also important in this respect, as well as in membrane performance, is pH. It should be kept in mind that multivalent ions such as calcium and magnesium are rejected to a greater degree than monovalent ions like sodium or chlorides. Higher-TDS waters require higher pump pressures.

If the source of influent water is surface water, frequent changes can be expected. Ground-water quality usually remains fairly constant; constant-quality raw water should be sought when possible.

Concentration Polarization.   Since the RO process takes place at the membrane, it can be expected that solute concentration will increase at the membrane wall. Reverse osmosis operation rejects a given percentage of the solute present at the membrane wall; accordingly, the concentration of solute at the membrane wall interferes with solute rejection. Film concentrations can be reduced by high velocities and turbulence enhancers. This is provided to a greater degree with spiral-wound vs. hollow-fiber design.

Bacteria-Algae.   Membranes can foul through direct bacterial attack on the membrane, growth on the membrane surface, or transport of *bacteria-algae* slime from other parts of the system. Where possible, chlorine is recommended for the control of microorganisms. Cellulose acetate membranes resist chlorine residuals up to 1.0 ppm. Polyamide membranes cannot tolerate more than trace amounts of chlorine. Under these conditions, nonoxidizing *biocides* should be used. When biocides other than chlorine are employed, it should be determined that they have no detrimental effect on the membranes. *Slime* deposits present on membrane surfaces are usually removed during cleaning operations with the use of *surfactants.*

Membrane Age.   All membranes deteriorate with age, regardless of operating conditions. An average life is two or three years. Monovalent ions will leak first, or the amount of permeate will decrease. Since the membranes have a limited life, there is no solution except to replace the membranes.

Operating Pressure.   The effect of increasing operating pressure is twofold. An initial result is to increase the amount of permeate. Since the amount of salt diffusion remains constant, the quality of the product water improves. On the other hand, increased pressure can compact the membrane wall. As a result, the quality of the permeate will improve, but the flow rate will decrease. In addition, the increased pressure can result in breaks in the membrane wall. Quality then deteriorates. Operating pressures recommended by the manufacturer should be followed.

Membrane Deterioration

Membranes are expensive and have a limited life. Accordingly, it is important that they be used under proper operating and maintenance conditions. The most common types of membranes are polyamide, cellulose acetate, and thin film composite. Development in membrane technology has been rapid in recent years. For this reason, recent information should be obtained when new installations are made. Common items to be considered are as follows:

Chlorine. While chlorine is an economical method of controlling bacteria, it is also a strong *oxidizing agent;* it reacts with many materials. Cellulose acetate will resist chlorine up to approximately 1.0 ppm free chlorine (Staff, *Industrial Water Engineering,* 1973). Polyamide and most thin film composite membranes have poor resistance to chlorine. A commonly recommended limit is 0.1 ppm free chlorine. Recently, a composite polysulfone membrane has been introduced that is resistant to chlorine up to 5 ppm free chlorine.

When recommended limits of chlorine are exceeded, the feedwater should be treated with a *reducing agent.* Sodium sulfite and metabisulfite are two chemicals often used for dechlorination. If biocides other than chlorine are employed, specific recommendations should be received from the supplier.

Temperature. Low temperatures reduce the production of permeate but do not damage the membrane. On the other hand, high temperatures can increase the rate of hydrolysis of membranes and shorten their operating life. The temperature operating limit for cellulose acetate membranes is 85°F. Recommended maximum operating temperature for polyamide membranes is 95°F. Thin film composite membranes are often recommended up to 113°F.

pH Control.   The life of the membranes is limited by the pH operating range. For cellulose acetate membranes, the operating pH range is 2.0–8.0, with a recommendation of 5.5. At 5.5 pH, calcium carbonate scale is avoided. Above pH 8.5, the cellulose acetate membranes will slowly deteriorate. Polyamide and thin film composite membranes will tolerate a wide pH range of 3–11.

Hydrolysis.   Hydrolysis, or chemical deterioration, of cellulose acetate occurs with time but is accelerated by high temperatures and high or low pH. Polyamide membranes do not hydrolyze but can fail from physical factors (Dupree, 1973). To extend the operating life of membranes, recommended temperatures, pressures, and pH should be followed.

As pressure is applied to a membrane, it slowly compacts. The amount of product falls off to a point at which the membrane must be replaced. Recommended back-pressure limits for permeate in a hollow-fiber system are 50 and 20 psig for a standard spiral-wound unit.

Case histories are recorded in which compaction has reduced output 30% in a year. Operating pressures designated by the manufacturer should be followed.

Bacteria.   Cellulose acetate membranes are sensitive to attack from bacteria. To prevent such occurrences, chlorine should be fed continuously, maintaining a free-chlorine residual of 0.5–1.0 ppm. Since polyamide membranes cannot tolerate chlorine, some other type of nonoxidizing biocide should be used for control. If bacteria are not controlled, the accumulation of slime in the membrane can shorten its life.

Part **Boilers and Auxiliaries**

In any steam-generating plant, the boiler itself represents the heart of the system. Depending on the size, pressure, intended use, etc., other auxiliaries may be present. The most important auxiliaries are the *deaerator, feedwater heater, economizer, superheater,* turbine, and condenser. Some low-pressure boilers have none of these auxiliaries, while large, high-pressure plants include all of them.

Boilers are of two general types: *fire tube* and *water tube*. Fire tube boilers are in much wider use but, in general, the pressure and size of the fire tube units are smaller than water tube boilers. The vast majority of the fire tube boilers are of the *packaged* design, that is, constructed at the boiler manufacturer's plant rather than field-erected. As the name implies, the combustion gases of a fire tube boiler pass through the tubes while the boiler water is on the outside. With an increase in pressure and capacity, the cylindrical wall thickness has to be increased. A point is reached at which it is impractical to increase the wall thickness further. Most fire tube boilers are limited to 250 psi, while their rating is usually less than 25,000 lb/h. Designs of fire tube boilers vary, but the vast majority of the fire tube units produced today are of the Scotch design. The Scotch boiler is a cylindrical furnace built inside the outer shell; the furnace area is surrounded by water. After passing through the furnace, the combustion gases reverse flow and pass through the tubes located largely above the furnace area. With very few exceptions, the tubes of fire tube boilers are not exposed to appreciable amounts of *radiant* heat. Also, the tubes are almost always surrounded by water. As a result, most tube *failures* in fire tube boilers do not involve high temperatures. When tube failure takes place in a fire tube boiler, it usually involves *pitting* from poor storage practices or insufficient oxygen removal during operation.

Circulation is not as well defined as in water tube boilers. As a result, scale can be a problem even though high radiant heat is absent. Virtually all larger, field-erected boilers and many industrial packaged boilers are of the water tube design. As the name implies, water passes through the inside of the tubes, and heat is applied on the outside. Basically, this is a safer design than the fire tube boiler since the amount of water involved in any failure is confined to the relatively small quantity present in the tube.

Many of the tubes in a water tube boiler are exposed to the radiant heat in the furnace area. As a result, tube failures usually involve overheating. High heat transfer also necessitates the use of high-quality feedwater to avoid the formation of scale. For the large majority of boilers, scale prevention primarily involves the removal of hardness. For high-pressure utility boilers, virtually pure makeup water is involved.

Although the boiler represents the major part of a steam-generating system, the auxiliaries are also important.

Deaerator. The deaerator is primarily a protective device for the boiler. It removes oxygen and free carbon dioxide from the feedwater before they have a chance to enter the boiler. A second benefit is the preheating of the makeup water to the boiler. Gas elimination is made possible by a combination of heating the incoming water and dispersing the water into very fine particles. An efficient deaerator should reduce oxygen to below 8 ppb. Most problems evolve from the failure of the deaerator to remove oxygen and carbon dioxide as designed.

Feedwater heater. Feedwater heaters are another means of increasing boiler efficiency. Sources commonly used to elevate the feedwater temperature are exhaust or extraction steam from turbines, engines, etc. Optimum performance on the steam side requires venting of noncondensable gases and maintaining of clean, noncorroded surfaces.

Exchangers are of the shell and tube design, with condensing steam on the shell side. Shell and tube heat exchangers are not restricted to boiler feedwater. They are used throughout industry to heat and cool many different products. The problems encountered in the use of steam on the shell side are similar, if not identical, regardless of the product on the tube side. One exception is leakage, where product contamination becomes involved.

Shell and tube exchangers are of several basic designs. The fixed tube sheet is the least expensive, but it suffers from thermal stresses between the tube and tube sheet. To alleviate this problem, *bellows* are often added. Floating heat exchangers are also often employed.

Economizer. Economizers are another means of increasing overall generation efficiency. This is done by passing combustion exhaust gases through a heat exchanger (economizer) containing feedwater. One estimate shows a 2% increase in boiler efficiency for every 25°F increase in feedwater temperature. Use of the economizer has expanded a great deal in recent years as the cost of fuel has increased.

Economizers are available in the form of both vertical and horizontal tubes. The vertical economizer is an integral part of the boiler; separate economizers consist of horizontal tubes. Water in economizers normally flows upward although some units operate in the opposite manner. The efficiency of an economizer increases as the inlet temperature decreases.

Water flow opposite the direction of the exhaust gas is considered to be the more efficient. Tubes usually range in size from 1–2 in. in diameter, and the ratio of total heat-transfer area is usually about 2 to 1 for boiler vs. economizer. In some cases, the economizer surface is approximately the same as the boiler. Since soot can collect on the outside of the tubes of economizers, they are normally fitted with soot blowers.

Superheaters.   Saturated steam is that state in which the temperature corresponds to its pressure. Moisture in suspension may or may not be present. Where it is not present, it is known as *dry saturated steam*. If the temperature of dry saturated steam is lowered, moisture will appear. If heat is added to dry saturated steam at the same pressure, the steam is said to be superheated. The temperature at the same pressure may then be lowered a certain degree without moisture appearing. This property is especially useful for turbine operation or engines, where an excessive amount of moisture can result in severe erosion damage. Superheated steam can also be transported over long distances with little heat loss.

Superheaters are divided into two main types: *convection* and *radiant*. The convection type is heated by the gas stream, while the radiant type is exposed to radiant heat in the furnace area. It is difficult to maintain temperature control with one type of superheater, and so combination superheaters are often used. Superheater temperatures increase in the convection type with load, while the reverse is true with radiant-type superheaters. The steam temperature changes in the superheater outlet are affected mainly by variations in steam load.

Turbines.   Turbines are machines that convert the potential energy of steam first into kinetic energy and then into mechanical energy. This is done by expanding steam through nozzles against blades on a rotor. Turbines are employed in many different types of operations, such as driving generators, centrifugal pumps, fans, and compressors.

There are two basic types of turbines: impulse and reaction. Impulse turbines make use only of high-velocity steam striking rotor blading. Pressure drop occurs only in stationary blades. When steam leaves a nozzle, a reactionary force is also created. When the turbine makes use of this force, it is known as a reaction turbine; expansion occurs in both the stationary and moving blades. Many turbines are a combination of impulse and reaction designs.

Turbines, in general, consist of rotors containing blades, a casing containing rotors, and stationary nozzles.

Condensers.   The larger the pressure drop through a turbine, the more work can be obtained through that unit. A condenser is used to remove the *latent heat* from the steam, converting the steam to water. This in turn creates a vacuum. Condensers are any type of equipment that converts vapor to liquid. Although any type of cooling can be used, water is most often employed. There are three common types of condensers employed but, for steam plant application, the surface condenser is more widely used. It is the only type considered in this book. The other two types are jet condensers and barometric condensers.

Surface condensers operate by passing cooling water through tubes while steam is condensed on the outside of the tubes. Heat is transferred through the tube walls. Although the operation is simple, design factors are quite important in assuring efficient operation. Gases, including air, are mixed with the steam as it enters the condenser. These gases must be removed without at the same time removing too much water vapor. Also, design provisions must be made to avoid subcooling the condensate. If this is not done, energy will be wasted in converting it back to steam.

4 Preboiler Auxiliaries

Faulty Deaerator Performance

The main purpose of deaerator operation is the removal of oxygen and free carbon dioxide; the deaerator (Fig. 4-1) also preheats the boiler feedwater preventing *thermal shocks* to the boiler(s).

Below Standard Gas Removal

Temperature.   Oxygen and free carbon dioxide are less soluble in water at elevated temperatures. In normal operation, the temperature of the deaerated water should be within 1–2°F of the saturated steam at the deaerator operating pressure.

If the deaerated water temperature is low, the cause should be investigated; the unit may be overloaded or underloaded. Spray nozzles may be defective or plugged. If trays are used, the trays may be broken or tipped. Steam-jet atomizing valves are sometimes used and may not be operating properly. Sufficient steam pressure may not be available at all times.

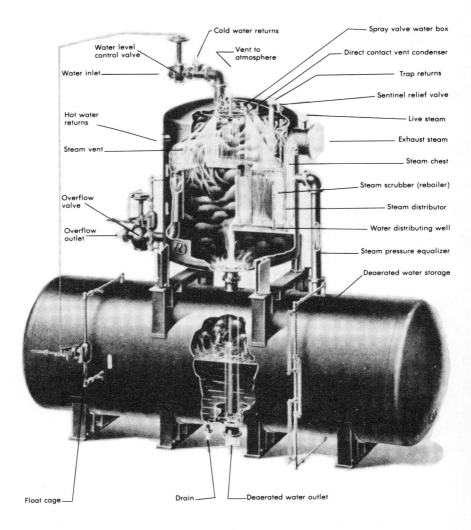

Cold water returns

Water level control valve

Vent to atmosphere

Water inlet

Hot water returns

Steam vent

Overflow valve

Overflow outlet

Spray valve water box

Direct contact vent condenser

Trap returns

Sentinel relief valve

Live steam

Exhaust steam

Steam chest

Steam scrubber (reboiler)

Steam distributor

Water distributing well

Steam pressure equalizer

Deaerated water storage

Float cage

Drain

Deaerated water outlet

Figure 4-1. Deaerator heater—spray type.

When the reason for low exit temperature is not readily apparent, the equipment manufacturer should be contacted.

Time. Needless to say, a certain amount of time is necessary to separate oxygen and free carbon dioxide from water. For this reason, the deaerator should not be overloaded. When it is not possible to avoid overloading a unit, it is all the more important that all other factors be correct.

Water Level. Changes from designed water levels should be avoided. High water levels can result in severe knocking. Damage to trays can result.

Turbulence. Turbulence brings gases to the surface and allows for their release. It also sweeps away the released gases from the water surface. Pressure-loaded nozzles assist in creating turbulence in the incoming water; trays also aid in this respect. Incoming steam is another factor contributing to turbulence. Atomization of water exposes a much larger surface than boiling.

Normal operation of the nozzles will cause turbulence. For this reason, only clear, nonscaling water should be added through the nozzles. Scale-forming water can also interfere with normal flow over deaerator trays; broken trays also decrease turbulence. Overall, when scaling waters cannot be avoided, spray-type deaerators are better suited than tray-type deaerators.

Thin Films. Thin films enable rapid transfer of gases to the surface, allowing quick release of the gases from the water. This permits the steam to sweep the oxygen and free carbon dioxide from the main body of water.

Trays in the deaerator and atomizing nozzles aid in creating thin films. Broken trays, deposits on trays, defective nozzle springs, or plugged nozzles all interfere with the formation of thin films. Tipped trays result in improper flow over the trays.

Venting. Venting allows the passage of the released gases to the atmosphere. The condensing steam plume from the vent is one indication that gases are being released. A plume of 2–3 ft represents normal operation; vent steam rates vary from about 0.05 to 0.14% of the rating of the deaerator.

The drains from an external vent condenser can be returned to the deaerator or discarded. Since these drains are saturated with oxygen and carbon dioxide, they are quite corrosive. In most cases, it is advisable to discard the drains.

Quite often, the tubes in the external vent condenser will corrode, allowing passage of raw water to the deaerator storage section. This water will be saturated with oxygen and free carbon dioxide. Water of this quality, of course, should not be allowed to return to the deaerator. Tube replacement should be made as soon as possible. If copper alloy tubes have been used, a change to stainless steel should be considered.

The conductivity of the vent condenser drains should be determined on a regular basis to detect any leaks. When no leaks are present, the conductivity will be low. The reverse will be true if leaks develop.

Load Variations. Low loads can often cause as many problems as high loads. When steam load requirements drop to a certain point, the steam jets or orifices do not operate effectively. The manufacturer's warranty will usually specify the minimum flow rate at which specified oxygen removal will occur. This also applies if the feedwater to the deaerator operates at too high a temperature. A temperature rise of at least 50°F in the deaerator over the incoming water is recommended.

Miscellaneous. Cold makeup water and hot condensate should not be mixed before entry into the deaerator. This mixture is quite corrosive since it contains water with dissolved oxygen at an elevated temperature. Separate lines should be used.

If cold makeup and hot condensate are mixed, the line

should be installed well before the deaerator and made of stainless steel.

A well-designed and operated deaerator should reduce oxygen to less than 8 ppb and remove all free carbon dioxide. Much of the carbon dioxide in water is combined as carbonate and bicarbonate. Most of this carbon dioxide is released in the boiler at higher temperature. Field oxygen analyzers are now available and should be used to check deaerator performance. The oxygen analyzer should be connected to a recorder so that the deaerator operation can be followed over a 24-h period. If high readings are obtained, the various operational aspects of the deaerator should be investigated.

Deaerator effluent low oxygen readings do not necessarily mean that the same-quality water is entering the boiler. Air quite often leaks into the boiler feed pump. Also, oxygen-saturated water is often added through a lantern ring to lubricate packing. In most cases, this water seal is not necessary; but the pump manufacturer should be contacted before the water addition is stopped at this point. If the seal must be used, it should be deaerated water, not raw water. A test for oxygen should be made on the feed pump discharge as well as the deaerator effluent.

It should be noted that a vibrating, noisy deaerator is usually a sign that the unit is not operating correctly. Mechanical problems are generally involved, and the manufacturer of the unit should be contacted. Any sign of *pitting corrosion* in the storage area detected during shutdown periods is another indication that the deaerator is not operating correctly.

Tray-type deaerators are better suited for varying loads or inlet water temperatures. Spray-type deaerators are best suited for scaling water.

Feedwater and Other Heat Exchangers— Corrosion and Deposits (Steam Side)

Extraction or exhaust steam is applied on the shell side of shell and tube heat exchangers to preheat boiler feedwater. However, a boiler feedwater heater is merely one use of a shell and tube exchanger. They are also employed in industry to heat and cool many different products. Where steam is used on the shell side, the problems encountered are similar, regardless of the liquid on the tube side. One major exception is leakage, where a product can contaminate the steam or the steam can enter the product side.

Although different problems exist on the shell (steam) and tube (feedwater) sides, they are interrelated. Any change in performance on one side affects the other. Usually, if the temperature and flow rate of the feedwater and the pressure of the inlet steam are the same as designed, then all other operating factors should remain constant. Any changes in pressure and temperature can indicate a possible problem.

The quality and operating conditions of the steam have a definite effect on the heat transfer as well as the life of the equipment. The operating engineer has no control over design of the heat-transfer equipment but, in many cases, he can modify the quality of the steam and the methods of operation.

When a feedwater heater fails to deliver the required feedwater temperature, a check of the steam pressure on the shell side should be made. Also, any accumulation of water in the shell should be noted. If these checks prove satisfactory, other factors should be investigated.

Corrosion

pH.   The pH of the condensing steam should be in the general range of 8.3–8.5. A pH of 6.0 or lower can result in the corrosion of most materials of construction used in heat exchangers. Neutralizing amines are commonly used to control the pH; in some cases, no pH adjustment is made, but filming amines are used to form a protective film throughout the steam condensate system. Deaerator performance should be reviewed to make certain that all free carbon dioxide is being removed.

High-pressure boilers use demineralized feedwater, which contains virtually no carbonate, but carbonate is present in low-and medium-pressure units. Breakdown of carbonate alkalinity is responsible for the formation of carbon dioxide and low-condensate pH. If alkalinity addition is required in boiler water, caustic soda can be used instead of soda ash. Caustic soda does not form carbon dioxide as a decomposition product. Various types of pretreatment can also be used to reduce carbonate alkalinity. (Refer to the material on carbon dioxide in Chapter 8.)

Ammonia. Ammonia will corrode copper-bearing alloys, resulting in stress corrosion cracking or grooving. Attack is aggravated by the presence of oxygen. Dosages of nitrogen containing water treatment chemicals such as neutralizing amines, filming amines, and hydrazine should be held to a minimum. Another source of ammonia is polluted makeup water.

Mechanically, performance of the deaerator should be checked so that ammonia is held to a minimum. Chemically, high-pH makeup water in a deaerator will result in more ammonia release than a unit operating at a medium-pH range such as 7.0–8.0.

Oxygen and Carbon Dioxide.   These two gases are present in most steam systems to some degree. Both gases intensify ammonia attack on copper alloys. Oxygen alone is the principal cause of pitting of steel, while carbon dioxide causes grooving.

Oxygen is usually held to a minimum by deaeration and the use of oxygen scavengers; only free carbon dioxide is removed in the deaerator. Various dealkalizing pretreatment methods assist in removing combined carbon dioxide such as bicarbonates and carbonates. Air leakage into any part of the steam condensate system will contribute to oxygen and carbon dioxide; the only solution is locating and repairing the leaks. During shutdown periods, *nitrogen blankets* can be used to prevent air in-leakage.

Venting. Proper venting of the steam side of a heat exchanger is necessary both to improve heat transfer and to reduce corrosion from gases such as oxygen, carbon dioxide, and ammonia. Venting can be continuous or intermittent, depending on the rate of gas accumulation. If the amount of noncondensable gases is high, the heat transfer will be reduced, and iron and/or copper levels will be increased.

Operating efficiency of a heat exchanger can be determined by observing the temperature of the saturated steam and the water leaving the exchanger. Effluent water temperature should be constant if steam and flow conditions are unchanged.

Vents should be inspected to make certain they are open; design should be such that there are no bends or restrictions in the vent lines. If operating under atmospheric pressure, the vent lines can lead to the condenser.

Impingement. Copper and some of its alloys are especially susceptible to *impingement* attack, which can also occur with mild steel. Often, this is a result of high steam velocity and the presence of moisture in the steam. Protection against this type of attack involves the installation of baffles in the heat exchanger or the covering of affected tubes with stainless clips. Stainless steel is usually not affected by this type of attack.

Vibration. Cross-flow high steam velocity is a common cause of vibration, but proper design of baffles, impingement plates, supports, etc., can hold vibrations to a minimum. U bends are more susceptible to vibrations than straight runs of tubing.

Vibration can result in tube failure or thinning at baffles or tube supports. These areas should be inspected during turnaround. Since the operating engineer has little control over design, the feedwater heater manufacturer should be contacted if damage is noted. Additionally, the exchanger should not be located close to vibrating equipment.

Exfoliation.   This type of corrosion occurs primarily in cupro-nickel alloys. It usually occurs in feedwater heaters of utility systems that operate on a peaking schedule. The entrance of oxygen during idle periods is evidently responsible. *Exfoliation* is characterized by the formation of black deposits that form as flakes, which then peel off. In time, the tubes will fail.

The easiest method for avoiding this type of corrosion is to maintain a small amount of steam or nitrogen pressure on the system, thereby preventing the entrance of air.

Leaks.   Where leaks are present, it is possible for the liquid (water or product) to contaminate the steam. Actual direction of the leakage (liquid to steam or steam to liquid) depends on several factors, including pressure.

The majority of leaks occur in the area of the tube sheets in shell and tube heat exchangers. Leaks can also result from gasket deterioration or failure to seat. Or they can result from tube corrosion or *fatigue* failures. Depending on the product on the liquid side, it is possible for a corrosive chemical to enter the steam side, with adverse results possible in most parts of the boiler system. Where this possibility exists, an automatic dump system should be installed in the condensate system. In some cases, it is advisable to dump all condensate at all times.

Salts—Caustic Soda.   Salts and caustic soda can carry over with steam and form deposits in the desuperheater section of heat exchangers. Corrosion can follow. (Refer to Types of Corrosion Encountered in this chapter.)

Deposits

Oil.   Oil is undesirable in that it can coat heat-exchanger sur-faces and interfere with heat transfer. Many plants use oil to lubricate reciprocating equipment. Cyclone or impingement separators can be used to hold the oil content in steam to a minimum. Condensate containing objectionable amounts of oil should be filtered or the condensate discarded. Aluminum sulfate plus caustic soda or soda ash is sometimes fed before an anthracite filter. Oil combines with the resulting aluminum hydroxide.

Salts—Caustic Soda.   Under normal operation, deposits are not a problem on the steam side of feedwater heaters. However, if carryover takes place, deposits can form in the de-superheater section of a heater. If caustic soda or chlorides are present, they can concentrate and cause severe corrosion problems. (Refer to Stress Corrosion Cracking, later in this chapter.) Iron and copper oxides are usually not present at this location. If they are in the generating system, they usually deposit earlier, before reaching the feedwater heater.

If no superheat is present on the shell side of a heat exchanger, deposits are unlikely. Where deposits form, the solution lies in taking the necessary steps to avoid carryover. Refer to Chapter 5.

Feedwater Heater—Corrosion and Deposits (Water Side)

Corrosion Related to Water Quality

The plant engineer has reasonable control of the chemistry of water on the tube side of feedwater heaters. Under normal operation, the feedwater to a high-pressure boiler represents a large percentage of condensate, plus high-quality makeup water. As a result, the corrosion problems that occur in high-pressure system feedwater heaters usually result from impurities that contaminate the feedwater. Feedwater in low- to medium-pressure systems is of lower quality, and a greater amount of corrosion is likely to occur.

In power plants, low-pressure heaters commonly use tubes of 304 stainless steel, copper-nickel alloys, and steel. Carbon steel tubes are most common in high-pressure heaters.

pH.   The pH of the feedwater should be held in the range of 8.5–9.0. This can be controlled in high-pressure systems with neutralizing amines such as *morpholine* or *cyclohexylamine*. In medium- or low-pressure systems, alkaline water treatment chemicals can be used. pH is involved in most forms of corrosion attack on feedwater heater tubes.

Oxygen and Carbon Dioxide.   Oxygen causes pitting corrosion of feedwater heater tubing, while carbon dioxide causes general attack when the feedwater is primarily condensate. Oxygen also contributes to velocity-related attacks on the inlet to feedwater tubes. Both gases can be controlled with proper deaeration.

Any remaining oxygen can be removed with *hydrazine* or sodium sulfite. Only free carbon dioxide is removed in the deaerator. Any carbon dioxide remaining in the boiler steam must be neutralized with neutralizing amines.

It is still possible for in-leakage of oxygen to take place after the deaerator at the feed pump. Accordingly, oxygen determination should be made on the water entering a feedwater heater.

In-Leakage.   In-leakage of raw water into the condenser can contaminate the feedwater. The feedwater conductivity should be monitored constantly.

If stainless steel tubes are used, it is important that the system be kept clean. In addition, chloride ion concentration should be held to a minimum.

Ammonia.   Although ammonia is at times used for pH control of condensate, it is corrosive to copper alloys. Breakdown of morpholine, cyclohexylamine, and hydrazine also contribute to ammonia formation. Where neutralizing amines or hydrazine are used, it is necessary to monitor for ammonia and copper. Overdoses of chemicals that contribute to ammonia formation should be avoided.

Deaerator performance should also be checked; ammonia removal will be more effective at high pH levels than at neutral or low levels.

Types of Corrosion Encountered

Designs, temperatures, and materials of construction of feedwater heaters vary. Accordingly, many types of corrosion are encountered. These include (1) general, (2) pitting, (3) stress, corrosion cracking, (4) impingement, (5) cavitation, and (6) crevice.

Since these classifications of corrosion are covered under cooling water corrosion problems, they are not repeated here. The reader is referred to Chapter 9.

Deposits

In high-pressure boiler systems, deposits are seldom a problem in feedwater heaters. Makeup is demineralized water, and condensate polishers remove any returning contaminants.

Where lower-pressure boilers are concerned, deposits usually result from faulty pretreatment or in-leakage. In such cases, the solution lies in correcting the faulty pretreatment or in taking steps to stop the leakage.

Economizer Corrosion and Deposits

Corrosion

Corrosion is the primary problem encountered with economizers on both the water and gas sides. Most corrosion occurs on the gas side through condensation of sulfuric acid but, as the title of this book suggests, only water side corrosion will be covered here.

The primary water side corrodent is oxygen, which can cause severe pitting. Unlike a boiler, where oxygen can be released in the steam drum, the oxygen in an economizer is trapped. Since no water treatment steps can be taken in an economizer, the answer to the problem is to remove all oxygen before the boiler feedwater enters the economizer. This means operating the deaerator efficiently and adding a sufficient amount of sodium sulfite or hydrazine to the deaerator storage. It is also possible that leakage of oxygen into the boiler feed pumps can take place.

Most economizers are of the horizontal type. Water can enter from the top or bottom. Water entering from the top is much more liable to trap air and initiate corrosion. It is also important in economizer corrosion control to maintain a pH of 8–9. At times, this is accomplished by recycling a small amount of boiler water. Most economizers are not of the steaming type. This is especially true of high-makeup boiler plants. As a general rule, the economizer water should be 40–50°F below the temperature of steam formation (Shields, 1961, pp. 274–282). Failure to keep economizer water within this temperature range can result in *water hammer*, with possible equipment damage.

Of probably greater importance in the corrosion of economizers is shutdown time. As with boilers, dry storage is preferred for long idle periods, while wet storage is preferred for short periods. For dry storage, all water should be drained or blown from the tubes. Heat should be applied to make certain all moisture is removed. Finally, the unit should be sealed and protected with a nitrogen blanket. For wet storage, deaerated water should be used: to this, about 100 ppm of hydrazine should be added.

High water temperatures (250°F) are advantageous for the economizer since they assist in preventing the condensation of sulfuric acid on the gas side. If possible, the pressure of the deaerator can be raised, or a heat exchanger can be used before the economizer. Adversely, if the temperatures are too high, steam may form. Water hammer can follow. Oxygen pitting is the major type of economizer corrosion. But where deposits form, it is also possible for caustic to concentrate underneath the deposits. Caustic gouging can result.

Deposits

Deposits are not as common a problem as corrosion in economizers. If calcium or magnesium salts are present, it is evident that proper water pretreatment is lacking. Pretreatment equipment and water treatment chemical controls should be checked. Deposits consisting of corrosion products such as iron or copper oxides are evidence that oxygen control is lacking. Deaerator operation should be reviewed. It is also advisable to check for iron or copper oxides in condensate return waters.

Some boilers employ caustic addition; with certain-quality feedwaters, the addition of caustic can make the water scale-forming, resulting in economizer deposits. Under these conditions, the caustic should be fed directly to the boiler drum.

Flashing of steam in an economizer can also result in scale formation. Steam formation is a design problem that should be reviewed with the engineering firm that installed the equipment.

Economizers are usually designed with large heating surfaces to absorb the relatively low-temperature gases. This results in lower water flow rates and increases the scale-forming tendencies. Even though the plant operator has no control over design, the importance of the operator's role in maintaining good-quality water should be clear.

5 Boiler Scale and Carryover Problems

Boiler Scale

Pretreatment steps reduce potential scale formation in boilers. Nevertheless, small amounts of scale-forming chemicals are still present in the feedwater. Major components of the scale-forming ions are calcium, magnesium, silica, and iron. At the elevated temperatures found in boilers, these ions will be involved in the formation of scale. In turn, scale can be the cause of tube failure or reduction of operating efficiency.

Scale can be prevented through the addition of chemicals that modify the scale-forming compounds. This can be accomplished by several methods. Foremost is the formation of insoluble compounds that have reduced scale-forming properties. Chief among these are the phosphates, which combine with calcium to form calcium phosphate (as *hydroxyapatite*). High OH alkalinities also assist in the formation of desirable insoluble magnesium hydroxide.

In addition to the formation of desirable insoluble chemicals, the scale-prevention program can be assisted through the addition of sludge-conditioner polymers. These compounds prevent scale formation through dispersing properties and/or interference in crystalline growth. The best known of these materials are the polyacrylates.

Finally, scale in boilers can be prevented through the formation of soluble complexes of the scale-forming ions. The best known of these chelating agents is *EDTA*. In spite of the addition of these compounds, scale in boilers is still a major problem.

Poor results using boiler water additives can occur through improper use. In addition, the quality of the feedwater can have an adverse effect on scale prevention. Certain operating conditions can also interfere with scale prevention.

Scale Related to Internal Treatment

Phosphate.   The addition of *phosphate* is the most widely used type of internal treatment for industrial boilers. The formation of calcium phosphate as hydroxyapatite is an effective method of handling calcium hardness. Improper use of phosphate can, however, result in scale formation. Usually, the phosphate residual is maintained in the range of 10–40 ppm. It is normally used with hydroxide alkalinity, sodium sulfite, and an organic sludge conditioner. Several errors in the use of phosphate can result in the formation of boiler scale. Although calcium phosphate (hydroxyapatite) is a desirable type of sludge, magnesium phosphate is not desirable. It forms a sticky sludge that is scale-forming. Magnesium phosphate usually develops because a high-phosphate/low-alkalinity ratio is maintained in the boiler water. Instead of forming a desirable magnesium hydroxide or silicate sludge, the scale-forming magnesium phosphate develops. Under low alkalinity conditions, adherent calcium phosphate deposits can form instead of a free-flowing hydroxyapatite sludge. The answer to this problem is to make certain that the phosphate residual is kept in a relatively low range (10–30 ppm) and that adequate alkalinity (OH) is maintained (Peters, 1980, p. 21). With most medium-pressure-range boilers, 200–400 ppm OH (as $CaCO_3$) is adequate. To confirm that magnesium phosphate is a problem, a deposit analysis should be made.

In the opposite direction, low phosphate residuals can also present a problem. Instead of the desirable hydroxyapatite being formed, calcium carbonate or calcium sulfate may precipitate. A quick field check on a low-phosphate problem is to add concentrated hydrochloric or sulfuric acid to a boiler deposit. If a rapid reaction takes place with *effervescence* of carbon dioxide, it indicates the presence of calcium carbonate, which would not be present if adequate phosphate treatment had been maintained. This simple field test should be confirmed with a deposit analysis.

Quite often, poor testing methods indicate that phosphate is present in the boiler water when the residual phosphate is actually low or absent. Before testing the boiler water sample, it should be filtered, using a slow filtering paper. Whatman 934 AH glass fiber filter membrane or equivalent is very effective for removing suspended phosphate. If the boiler water is not filtered, the test will include suspended calcium phosphate as well as the residual phosphate in solution.

If at all possible, phosphate treatment should be added directly to the boiler drum. When orthophosphate is added to the feed lines, the lower alkalinity often results in the formation of calcium phosphate scale in the feed lines or in the cold section of the boiler. If phosphate has to be added to the feed lines, a polyphosphate should be employed instead of orthophosphate. Most of the polyphosphate will not revert to orthophosphate until it reaches the boiler drum. Accordingly, scale formation in the feed lines may be averted.

Carbonate Cycle.   About 20 years ago, virtually all industrial boilers used phosphate to precipitate calcium hardness. When synthetic polymers were introduced, it was found that the new sludge conditioners were capable of conditioning calcium carbonate. The dispersed calcium carbonate formed free-flowing sludge instead of scale. The addition of phosphate was eliminated.

Problems with the *carbonate cycle* treatment result from a misapplication of the method. The carbonate cycle method should be used only with boilers operating at low pressure, usually less than 200 psi. At higher pressures, the carbonates decompose rapidly. As a result, there is an insufficient amount present to react with calcium hardness entering the boiler. A carbonate excess of at least 100 ppm ($CaCO_3$) should be present. This is determined by calculating 2 X (M-P) alkalinity.

If it is impossible to maintain the required carbonate excess, a change should be made to another treatment method. Phosphate or chelant treatment methods should be considered. In a few cases, the improper use of a sludge conditioner, such as tannin or sulfonated lignin, results in scale formation; the carbonate cycle treatment should be used only with the newer synthetic polymers.

Chelants. Since the *chelants* EDTA and *NTA* are designed to keep precipitates from forming, it might be asked how deposits can be a problem. Still, under certain conditions, they can form. Ferric iron, for example, is not strongly chelated by EDTA or NTA under boiler water conditions.

Any iron oxide (Fe_2O_3) entering the feedwater will not be solubilized by the chelants. When this situation exists, an iron-dispersing polymer should be used with the chelant. Testing for free-chelant residual should be made on the feedwater, not the boiler water. Free EDTA decomposes under boiler water conditions; NTA is more stable. If insufficient chelant is added, calcium sulfate, calcium carbonate, or other deposits can form.

Chelants should not be used in boiler systems in which the feedwater hardness is erratic. The treated makeup water hardness should be consistently low.

In some cases, the use of a chelant has been supplemented with phosphate. The reasoning behind this method is that phosphate will form desirable calcium phosphate (hydroxyapatite) sludge if insufficient chelant is added to a system. However, there is dispute concerning the effectiveness of this method. Some studies indicate that calcium phosphate will precipitate in spite of the presence of a chelate. If this method is used, the phosphate should be added directly to the boiler drum, whereas the chelant should be added to the discharge side of the boiler feedwater pump.

Hydroxide and silicate ions will also compete with chelant for the magnesium ion (Stephans and Walker, 1973). As a result, magnesium hydroxide or magnesium silicate deposits or sludge may form even though chelants are present. Magnesium hydroxides or magnesium silicate usually develop only when high hydroxide or silicates are present. Accordingly, if these deposits are present, steps should be taken to reduce one or both of these ions.

EDTA forms stronger bonds than NTA with calcium and magnesium. If deposits are formed and NTA is used, a change to EDTA should be considered. In all cases in which deposits form, an analysis should be made to determine the type of problem and indicate what corrective action should be taken.

Coordinated Phosphate. The coordinated phosphate treatment and its variation, "congruent control", are designed for high-pressure boilers. They depend on the *buffering* action of phosphates to prevent rapid buildup of caustic alkalinity. The coordinated phosphate treatment is seldom the cause of deposits if it is correctly used. The treatment is designed only for high-pressure boilers; it also requires very tight control on the quality of feedwater and control of phosphate and pH. Only demineralized makeup should be used. This type of treatment should not be used in industrial boilers operating at less than 1000 psi. Below this pressure, the control requirements in the caustic boiler water are not so stringent, making the coordinate phosphate method unnecessary.

Peaking plants in utility service often maintain skeleton crews. Under these conditions, poor control of the coordinate phosphate program may result. This method should not be employed unless the boiler operates in the pressure range that cannot tolerate hydroxide alkalinity. This type of internal treatment should not be used unless excellent water treatment controls are maintained.

Sodium Sulfite.   An oxygen scavenger, sodium sulfite, is normally not considered in relation to boiler-scale formation. There are exceptions. It has been known for many years that sodium sulfite decomposes in high-pressure boilers (900+ psi). Besides contributing to the formation of sulfur dioxide in the steam condensate system, sulfides of iron and copper can also form. The deposits can possibly form the nuclei for further scale formation.

There is also evidence that sodium sulfite (Limpert and Schroeder, 1978) can decompose to form cuprous and iron sulfide in medium-pressure water tube boilers. There the sulfides form in only very small amounts. The formation usually takes place only at the metal/scale interface; the area of sulfide formation is the furnace area, where high radiant heat prevails. The presence of sulfide can be checked by submerging a deposit in Lugol's solution (for preparation, see below). Effervescence of nitrogen gas will confirm that the deposit contains sulfide. As previously stated, cuprous sulfide or ferrous sulfide can serve as the nucleus for further scale development.

Sulfite decomposition will usually take place when oil or coal are used as fuel; both fuels produce a large amount of radiant heat. Where sulfide scale is indicated, the easiest solution is to discontinue the use of sodium sulfite and change to hydrazine. Hydrazine does not form any scaling products in boiler water.

Sodium Azide-Iodine Test for Sulfide

1. Lugol's solution: Dissolve 15 g of iodine and 30 g of KI in as small an amount as possible of demineralized water. Dilute to 100 ml.

2. Dilute 10 g of sodium azide in a small amount of de-
    ionized water.

3. Combine the two solutions above, and dilute to 500 ml. In
    performing the test, grind the sample to a fine powder.
    Place the sample in a test tube. Add the solution, and
    observe for effervescence. If effervescence occurs, sul-
    fides are present.

Polyacrylate.   Polyacrylates and methacrylates are the most
widely used of the synthetic sludge conditioners. When
polyacrylates were introduced, the molecular weight of the
polyacrylates was approximately 90,000; instead of dispersing
sludge, the polymer acted as a flocculant. If circulation was
poor, scale resulted. In the following years, the molecular
weight was reduced to the point at which most commercial
polyacrylates used in boiler water treatment now employ a
molecular weight between 1000–5000. The dispersing properties appear to be optimum in this range.

In evaluating scale problems, the performance of the polymer can be an important factor. If a scale analysis shows a large
percentage of hydroxyapatite (calcium phosphate), the sludge-
conditioning program should be investigated. Hydroxyapatite
is a desirable type of sludge. If it is a major part of a deposit, it
may mean that the polyacrylate is not performing its function.
In evaluating the polyacrylate, the investigator should determine the molecular weight of the product used. Also, the
amount of active polymer in the proprietary product should be
known. About 8–15 ppm of the active polymer is a normal
range in boiler water.

The quality of commercial polyacrylates varies considerably.
Accordingly, changes in the product should not be made merely because one material costs a few cents less than another.

The point of addition is also a factor in the performance of
polyacrylate. Like sodium sulfite, the polyacrylates should be
added as far back in the boiler system as possible. In most
cases, this will be the storage area of the deaerator.

Polyacrylates appear to perform best in the high-heat-transfer area of a boiler. They do not appear to perform as well in the colder section, such as downcomers or *mud drums* (Culsa, 1973). Combining the polyacrylates with other types of polymers appears to solve this problem. Most polymer sludge conditioners, including polyacrylates, appear to perform to the highest degree at higher alkalinities. Polyacrylates are also not especially effective in dispersing iron oxide deposits. Again, combining the polyacrylate with a specific iron *dispersant* has proved a desirable combination.

Often, polyacrylates on either the phosphate or carbonate cycle are not able to produce satisfactory results. Should this occur, a change to a chelant type of treatment is recommended. (Refer to Uneven Firing in Multiburner Boilers, page 93.)

Tannins and Sulfonated Lignins.   Tannins and sulfonated lignins are sometimes referred to as natural sludge conditioners. They were in wide use before the advent of the synthetic sludge conditioners. Their use at present is restricted to low-pressure boilers; at high pressures, they carbonize in the high-heat-transfer areas. The quality control of these products also varies considerably. If a deposit problem exists when a tannin or sulfonated lignin is used, a change to a newer synthetic polymer should be considered.

Sulfonated Styrene-Maleic Anhydride Copolymer.   This synthetic sludge conditioner has been coming into wider use in recent years. It is excellent as an iron oxide transport agent. It appears to be a better dispersing agent than a *crystal* modifier. In most applications, the optimum results are obtained when it is combined with polyacrylate or some other polymer.

If a deposit problem exists during the use of this material, it should first be determined if it is combined with another polymer. If not, a change to a combination should be considered. If a combination product is being used and a scale problem exists, the dosage should be checked. From 8–15 ppm of total active polymer should be present in the boiler water.

If the above aspects are in order and a scale problem still exists, the solution probably lies in something other than the choice of sludge conditioner.

Filming Amines. This product is covered more fully under condensate line treatments. Suffice it to state here that the use of filming amines has resulted in the formation of sticky balls in the boiler. Usually, these balls are the result of the filming amine combining with iron oxide in the condensate lines. The subject is covered in greater detail in Chapter 8.

Chromate and Nitrite. Chromates or nitrites are frequently used in low-pressure heating boilers. Boilers of that type usually operate with a high percentage of condensate return water. Scale usually results when the products are misapplied; they are intended to be corrosion *inhibitors,* not scale inhibitors. When scale appears, it usually means that more makeup water is entering the system than was intended. If the percentage of makeup cannot be reduced, a change to a different type of water treatment program should be undertaken.

Scale can also result if chromates are used in boilers operating above 200 psi. The chromates tend to decompose into anhydrous chromic oxide and sodium hydroxide. In addition to scale forming, the corrosion-inhibiting effect of chromate is lost. Also, chromates should not be used along with organic sludge conditioners, since they may tend to oxidize the sludge conditioners.

Changing Water Conditioning Program. Frequently, heavy deposits are found in a boiler, and a change is made to a different type of program. Before a change is made, the boiler should be acidized to remove existing deposits. If a change is made without acidizing, the existing scale may break loose and plug some tubes. Overheating and failure of tubes can result; this is primarily a problem with water tube boilers.

Phosphonates.   In recent years, the phosphonates have come into use as scale-control agents. Diphosphonate (HEDP) appears to be the only phosphonate in general use. One major drawback to the use of AMP phosphonate is that it interferes with the sodium sulfite/oxygen reaction; HEDP does not exhibit this limitation. The performance of HEDP has been somewhat erratic. The product appears to be more effective at low vs. high concentrations or when used in combination with a polymer such as polyacrylate. In combination with a polymer, phosphonate is effective at both high and low dosages. Unlike a program employing a polyacrylate sludge conditioner only, the combination polymer-phosphonate program is effective in colder areas of boilers as well as in the high-heat-transfer areas. HEDP has also proved an excellent iron oxide scale-control agent.

If a scale problem develops with the use of phosphonate, it should be determined if the phosphonate is being used alone or in combination with a polymer. If alone, a change should be made to a combination treatment (Good, 1961). The point of addition should also be observed. As with chelants, oxygen has an adverse effect on phosphonates. After deaeration and addition of sulfite, oxygen should be introduced only on the discharge side of the boiler feed pump.

Some decomposition of phosphonates takes place with long retention time. For this reason, phosphonates may not be appropriate for heating boilers using little makeup or for boilers operating at high cycles of concentration.

Caustic Soda and Soda Ash.   High-pressure boilers operate in the virtual absence of hydroxide alkalinity. However, the vast majority of operating boilers in the low- and medium-pressure range depend on the presence of hydroxide alkalinity to assist in the prevention of scale and corrosion. When sufficient hydroxide alkalinity is present in boiler water, the phosphate precipitates as hydroxyapatite. Magnesium hydroxide or magnesium silicate also precipitates. These end products tend to be non-scale-forming. On the other hand, if hydroxide alkalinity is lacking, tricalcium phosphate and magnesium phosphate are formed. Both are scale-forming.

Hydroxide alkalinity can result from the direct addition of caustic soda or soda ash. At boiler temperatures, soda ash decomposes to form caustic soda. Hydroxide alkalinity can also originate from the alkalinity present in the raw water. Refer to Table 5-1 for recommended boiler water concentrations of hydroxide alkalinity. Hydroxide (OH) alkalinity in a boiler is reduced by an increase in feedwater hardness, acidic contaminants in the condensate, or raw-water makeup. Ammonium salts entering a boiler break down to form acid plus ammonia. The ammonia leaves the boiler while the remaining acid reduces the alkalinity.

If salt from a poorly rinsed ion-exchange softener enters a boiler, the total dissolved solids (TDS) will increase. If the operator then increases the blowdown rate, the hydroxide alkalinity will be reduced.

In all cases, corrective action should be taken to maintain the hydroxide alkalinity in the proper range.

Scale Related to Feedwater Quality

As a result of elevated temperatures, certain salts present in boiler water decrease in solubility. Included are calcium carbonate, calcium sulfate, magnesium hydroxide, and various silicates. A major portion of water treatment is devoted to devising methods for either preventing the salts from forming or inhibiting them from forming scale.

Unfortunately, in an attempt to achieve these conditions, other problems often develop.

Hardness—Calcium and Magnesium.   For the vast majority of boilers, some type of pretreatment is employed. This might be a type of ion-exchange unit, hot- or cold-lime/soda treater, or some combination of these units. An increase in feedwater hardness usually means that a change has taken place in the

Table 5-1. Guidelines for Water Quality in Industrial Water Tube Boilers for Reliable Continuous Operation

Drum Pressure (psig)	Boiler Feedwater			Boiler Water		
	Iron (ppm Fe)	Copper (ppm Cu)	Total Hardness (ppm CaCO$_3$)	Silica (ppm SiO$_2$)	Total Alkalinity[a] (ppm CaCO$_3$)	Specific Conductance ($\mu\Omega$/cm)
0– 300	0.100	0.050	0.300	150	700[b]	7000
101– 450	0.050	0.025	0.300	90	600[b]	6000
451– 600	0.030	0.020	0.200	40	500[b]	5000
601– 750	0.025	0.020	0.200	30	400[b]	4000
751– 900	0.020	0.015	0.100	20	300[b]	3000
901–1000	0.020	0.015	0.050	8	200[b]	2000
0–1500	0.010	0.010	0.0	2	0[c]	150
0–2000	0.010	0.010	0.0	1	0[c]	100

[a]Minimum level of OH alkalinity in boilers below 1000 psi must be individually specified with regard to silica solubility and other components of internal treatment.
[b]Alkalinity not to exceed 10% of specific conductance.
[c]Zero in these cases refers to free sodium or potassium hydroxide alkalinity. Some small variable amount of total alkalinity will be present and measurable with the assumed congruent control or volatile treatment employed at these high pressure ranges.
Source: Presented at International Water Conference, 1975 by David E. Simon, NUS Corp.

pretreatment operation, resulting in undesirable hardness. Condensate can also become contaminated with hardness.

As a first step toward a solution, an investigation should be made to determine if changes have taken place in hardness in the treated makeup water or the condensate return water. If no previous records were maintained, tests should be made and the results compared against recommended standards for the boiler pressure employed. Examine Table 5-1 for recommended standards. If these standards are being exceeded, steps should be taken to bring the boiler controls into line. If the investigation indicates, for example, that increased or excessive hardness is coming from the pretreatment equipment effluent, then refer to Part I, Pretreatment Performance Problems.

Preliminary investigation may show that the increased hardness is coming from the condensate rather than the makeup treated water. If so, refer to Chapter 8, which covers the various problems that develop in condensate systems.

There are a significant number of boilers using untreated makeup water. For the most part, these boilers operate in the low-pressure range (10–100 psi) with normal amounts of makeup. The raw-water makeup for these systems is usually low in hardness, TDS, etc. A second group of boilers may employ raw-water makeup with moderate amounts of hardness but may be used only for heating. Under these conditions, very little makeup water is employed. The potential for scale formation is small; feedwater consists of close to 100% condensate.

The potential for scale is greater in the first system, which uses a low-hardness feedwater. As an initial step in investigating a scale problem in this type of system, it should be determined if there has been any change in the raw-water makeup quality. If the source is ground water, the composition normally does not vary to a great degree. On the other hand, a change may have been made in the well source. If the makeup is a municipal treated water, the water department can advise if any change has taken place.

If an adverse change in raw-water makeup results in boiler scale, two solutions are available. A zeolite softener can be installed to provide soft water. This involves capital expenditure. If this is not feasible, changes must be considered in the internal treatment program. These changes would be beyond normal recommended practice. One change to be considered is lowering the TDS of the boiler water, which acts as an indirect method of lowering suspended solids and also reduces the scaling potential. Another change that might be considered is increasing the frequency of the bottom blow. This assists in removing additional suspended solids. Increasing the hydroxide (OH) alkalinity reduces additional scale potential. Up to a boiler pressure of about 250 psi, an OH alkalinity of 600–800 ppm is not objectionable if foaming does not occur. If a polymer sludge conditioner is not being employed, it should be. Where a polymer is used, an increase in dosage assists in preventing scale formation.

As mentioned, the other types of boilers that might use raw-water makeup are those units that are devoted almost exclusively to heating. This type of boiler typically uses close to 100% condensate return as feedwater. Since the boiler is low-pressure and consumes virtually no makeup, almost any quality of makeup can be employed. Accordingly, if scale appears in this type of boiler, the quality of the feedwater is not the primary cause of the problem. Most often, it means that condensate is being lost and that the designed closed heating system is actually not closed. This type of boiler is often treated with chromate or nitrite. These chemicals are corrosion inhibitors and are not designed for scale prevention.

If the losses in the condensate return cannot be stopped, the boiler should be treated as a process boiler. The installation of pretreatment equipment may be required. Chromate or nitrite treatment should be discontinued and replaced with recommended treatments for hardness and oxygen removal.

Silicates.   In all cases in which deposit problems are involved, an analysis should be made of the scale.

Many analyses indicate the presence of silica; usually, the analysis shows combinations of silica with aluminum, iron, magnesium, or calcium. One compound, serpentine $(3MgO.2SiO_2.2H_2O)$ is a desirable type of sludge. When found in a deposit, it usually is mixed with one or more scale-forming compounds. Many other silica compounds are scale-forming (Hamer, Jackson, and Thurston, 1961, p. 117). These include

- analcite $(Na_2O.Al_2O_34SiO_2.2H_2O)$

- acmite $(Na_2O.Fe_2O_34SiO_2)$

- xonotlite $(5CaO.5SiO_2.H_2O)$

- pectolite $(Na_2O.4CaO.6SiO_2.H_2O)$

Acmite and analcite tend to form in high-pressure boilers (McCoy, 1981, p. 81).

Polymer sludge conditioners do little to prevent silica scales from forming. If appreciable amounts of silica are found in boiler deposits, the following corrective steps can be taken. First, the silica concentration should be kept within the ranges shown in Table 5-1. In some plant situations, of course, it is not possible to keep silica within the recommended ranges. In these cases, the OH alkalinity in the boiler should be increased. This approach can be used only with low- to medium-pressure boilers. Where high-pressure boilers are involved, steps must be taken to assure that the boiler water silica levels are not exceeded.

As with boiler hardness, if feedwater silica levels are exceeded, the makeup and condensate should be checked. For possible solutions to the high silica readings, refer to Part I, Pretreatment Performance Problems, as well as to Chapter 8, Condensers.

Silica in boiler deposits is usually combined with other constituents such as calcium, aluminum, and iron. The scale problem can be alleviated by maintaining close control of calcium, aluminum, and iron as well as silica.

Silica vaporization and carryover are also involved in turbine and superheater deposits. These problems are treated under the appropriate headings.

Iron (Oxides).   Iron in any of its *oxide* or complex forms is undesirable in boiler water. It is very difficult to disperse, so that it can be removed through the continuous or bottom blow-down lines.

Iron in its various forms can originate in the raw-water makeup or condensate return water or it can form directly in the boiler as a result of corrosion (Murphy, 1979). Unlike precipitation salts, most iron oxide originates outside the boiler. It does not concentrate in the boiler, and it tends to collect in stagnant areas.

Some steps can be taken to prevent the deposition of iron in boilers. If sodium sulfite is being used as an oxygen scavenger, a change to hydrazine can be recommended. Hydrazine will reduce the red oxide (Fe_2O_3) to the black magnetic oxide (Fe_3O_4). This oxide is somewhat easier to disperse in boiler water. Another recommendation would be to add a specific iron dispersant, such as the diphosphonate HEDP or sulfonated styrene maleic anhydride copolymer. Neither of these products should be used alone. Their performance is much improved if they are combined with a low-molecular-weight polyacrylate.

Among the older type of iron dispersants, lignin sulfonates are also recommended. However, their performance in dispersing iron is not as effective as the newer synthetic sludge conditioners. Needless to say, the most effective method of avoiding iron scale problems is to keep iron salts out of the boiler water. If iron oxide is formed as a result of boiler corrosion, corrective steps should be taken to stop the corrosion. This subject is covered in Chapter 6.

As mentioned, two other sources of boiler iron deposits are condensate and pretreated makeup. Control of iron in these areas is covered in Part I, Pretreatment Performance Problems and in Chapter 8, Condensers.

Oil. Although not a natural constituent of boiler water, oil can frequently enter a system through leaks in a condenser or other heat exchanger or through the lubrication of steam-driven reciprocating equipment. Whatever the source, the presence of oil in boiler water is undesirable. Oil can act as a binder to form scale. In high-heat-transfer areas, oil can carbonize and further contribute to the formation of scale.

Foaming is one indication of oil in boiler water. Its presence can be confirmed by first shaking a bottle or flask containing boiler water. If oil is present, foam will result. A small amount of powdered activated carbon should then be added. This time, little or no foam will appear if the foaming is caused by oil. Many other *organic compounds* react in the same manner.

If a sudden, unexpected slug of oil enters a boiler, severe foaming may result. Corrective steps must be taken immediately. Antifoam compound should be added or increased. The continuous blowdown rate should be increased to remove the oil from the boiler as quickly as possible. Temporarily, hardness should be allowed to enter the boiler. If phosphate treatment is being used, it should be discontinued until the oil is removed. Phosphate sludge forms scale with oil. In its place, the carbonate cycle should be used to absorb the oil. The system should be returned to normal operation when the oil is removed.

Often, oil in boiler water originates in the condensate. This contaminated condensate should be directed to waste until the source of the oil is determined and corrective steps taken. Taking a longer view, if oil leaks can be expected to occur frequently, consideration should be given to the installation of an oil monitor in combination with an automatic dump valve, so that any oil leaks in the condensate can be directed away from the boiler. Consideration can also be given to the installation of a centrifugal oil separator and/or filter.

A simple test for the source of oil consists of placing flakes of camphor on the surface of the water sample. If oil is present, the particles will remain motionless. On the other hand, if no oil is present, the camphor particles will move rapidly about the surface. The limitation of this test is that it is too sensitive for field use. Mere traces of oil will stop the movement of the particles. Accordingly, the test will indicate positive for a system in which traces of oil are present but no actual oil problem exists. The chief value of the test lies in clearing suspected areas. If the camphor test for oil is negative, one can be fairly certain oil is not a problem in that area.

Sulfate (Calcium Sulfate).   If a scale analysis shows the presence of sulfate, it is most often present as calcium sulfate (anhydrite). Calcium sulfate will form, of course, only if both calcium and sulfate are present. In addition, it will form only if scale inhibitors (phosphate, carbonate cycle, or chelant) are absent. Other than DI water makeup for high-pressure boilers or reverse osmosis treatment, there is no low-cost method of removing sulfates. The answer to the problem lies in pretreating properly to remove hardness and maintaining sufficient amounts of internal scale inhibitors.

Copper (Oxides).   There is some debate regarding the role of copper in high-pressure boiler-tube corrosion. Conversely, there is agreement that, where scale is concerned, no substance, including copper, is desirable. Corrective steps mainly involve keeping copper out of the boiler water. Refer to Condensate Systems Contaminants in Chapter 8.

Scale Related to Operating Conditions

Bottom Blowdown.   Heavy sludge tends to accumulate in the downcomer tubes or mud drum in water tube boilers. As expected, it will also settle in the lower part of fire tube boilers. If steps are not taken to physically remove this sludge, scale problems can be expected to develop eventually. The mechanical solution to this problem is to initiate a program of bottom blows. If, for example, one bottom blow per day is being performed, consideration can be given to increasing the number to two or three. The bottom blow should consist of one quick opening and closing of the manual blowdown valve. An extended opening can result in interference with the boiler's circulation.

Many boiler manufacturers do not recommend the blowing of the mud drum. This is especially true where high-pressure boilers are involved. If the boiler manufacturer advises against the use of the bottom blow, other means must be taken to reduce the amount of sludge or improve the method of handling the sludge. The sludge potential can be lowered by reducing the cycles of concentration through increased surface blow.

Other scale-reducing steps in this section can be taken to limit the amount of sludge formed.

Chemical Feed, Boiler Feed, and Surface Blowdown Lines.
Normally, a boiler manufacturer places these lines in their proper position. Nevertheless, field changes are often made, with adverse results. In water tube boilers, the feedwater pipe and chemical feed lines should be located so that they do not short-circuit to the continuous blowdown line. In a water tube boiler, the feedwater line is frequently located about $\frac{1}{4}$ diameter of the steam drum above the bottom surface of the drum.

The continuous blowdown line is used to remove suspended solids and maintain control of dissolved solids. It is often located about $\frac{1}{2}$ in. below the normal operating level. The chemical feed line should be located close to the boiler feedwater line to assure thorough mixing and reaction. The chemical feed line should also be located so that it mixes with the feedwater after the feedwater mixes with the alkaline boiler water. During boiler inspection, all pipe perforations should be inspected to make certain they are clear. The lines, too, should be clear and free from scale. Proper location of the three internal lines, and keeping these lines open and free-flowing, helps to maintain deposit-free boiler tubes.

Changes in Fuel.   Where physical factors are involved in scale formation, the causative agents are usually overheating and/or poor circulation. One major factor in overheating is a change in fuel. As an example, a change from gas to oil results in higher radiant heat being released in the furnace area. The chances of scale formation increase. A change in fuel type should not be made without consulting the manufacturer. Certain internal treatment chemicals and water conditions promote scale formation in the high-heat-transfer areas.

Low Water Level.   Low water levels in the boiler steam drum can result in steam being drawn into the downcomer tubes. This decreases the density of the water in that area (Firman, 1975) which, in turn, can have an adverse effect on circulation, resulting in possible scale formation. Every effort should be made to maintain proper water level.

Nearly Horizontal Tubes.   As might be expected, the circulation of boiler water in nearly horizontal tubes can present a circulation problem; this is especially true at low loads (Peters, 1980, pp. 16–26). Suspended solids can drop out in this area. Also stratification can occur, with steam on the top of the tube and boiler water on the lower part; splashing can then result in the formation of deposits. Since the operating engineer cannot change the design of the boiler, it is especially important that proper water treatment be maintained.

Cycling Operations.   Continuous, steady production of steam represents the optimum condition for boiler operation. Shutdowns and sudden changes in load present special problems.

Cycling boilers operate under an on/off schedule. For example, the boiler may operate only during the daytime hours. Or the plant may operate for five days and then shut down on weekends. Under these conditions, corrosion is the main problem, but scaling can also develop. When a boiler goes off-line, any suspended solids settle on the tube surface. They may not be dispersed when the boiler resumes operation.

Cycling operation can also change circulation patterns, sometimes resulting in the formation of scale. Unfortunately, boiler operations usually cannot be changed without considering plant requirements. Under the circumstances, the only preventive step that can be taken is to maintain excellent water treatment conditioning. This includes maintaining suspended solids as low as conditions permit; scale may not be eliminated, but it can be held to a minimum.

Cycles of Concentration—Total Dissolved Solids (Low and High).   In recent years, in an effort to conserve energy, there has been a tendency to increase the total dissolved solids in the boiler water. In individual plants, this may or may not be a rational move. Any increase in TDS also increases the tendency of the boiler water to form scale. Of even greater importance than the dissolved solids are the cycles of concentration based on the feedwater.

Boilers act as a sink for contaminants returning in the condensate, one chief impurity being iron oxide. To assist in preventing scale, a limit should be placed on the cycles of concentration. The limit on the number of cycles varies with each system but, as a general rule, the cycles should not exceed 50, or 2%, blowdown. This applies to high-pressure boilers. In medium- and low-pressure boilers, factors like silica and alkalinity usually limit the cycles to a much lower figure; 10–25 are common. Since the dissolved solids and cycles of concentration are controlled by the continuous surface blowdown, this line should be checked at regular intervals to make certain it is open. A touch of the hand to determine if the line is hot or cold is usually sufficient.

Occasionally, low cycles of concentration and/or low TDS occur in a boiler. Only indirectly can low cycles or TDS lead to increased boiler scale. If the low conditions are caused by a leaky blowdown valve, for example, more makeup water is required. This may exceed the limits of the pretreatment, enabling hardness to enter the boiler. More often, the low cycles merely represent inefficient operation. A leaky bottom blowdown valve can be detected by noting if the line is hot or at ambient temperature.

Low boiler water cycles or TDS can also result from excessive foaming or priming. This condition can cause deposits in the superheater tubes. Foaming and priming can usually be detected by violent swings in the boiler water level. The TDS of the condensate will also increase. Refer to Boiler Priming and Foaming at the end of this chapter.

Low boiler water cycles and/or TDS can also be caused by a change in the raw-water quality or a greater percentage of condensate return. Either condition is not undesirable. The cycles of concentration will not decrease, but the TDS of the boiler water will fall. Higher than normal total dissolved solids are sometimes encountered at times when there has been no apparent change in operation. High TDS can indirectly increase the tendency to scale since suspended or scale-forming materials may also increase. When an increase in TDS is unaccounted for, the following items should be investigated:

1. Continuous or bottom blowdown may be blocked. This can be checked by feeling if the line is hot during blowdown operation.

2. Raw-water makeup may have changed quality. Compare the analysis of present raw water with older analyses.

3. Ionized contamination may be returning with the condensate. This can be determined with a TDS meter.

4. The ion-exchange softener may not have been rinsed completely. Determine by analyzing for chlorides in boiler water. This will indicate if chlorides are abnormally high when compared to makeup water.

5. A smaller percentage of condensate may be returning to the boiler. This may result from leaks, deliberate dumping of condensate, etc. Percentage of condensate can be determined by checking the TDS of makeup vs. TDS of feedwater.

$$\% \text{ Condensate} = \frac{\text{TDS makeup} - \text{TDS feed}}{\text{TDS makeup}} \times 100$$

Changes in Operating Pressure. A change in operating pressure can also affect circulation. An increase in pressure can result in scale formation because there is less movement of water through the tubes.

When pressure is reduced, the specific volume (ft^3/lb) of the steam/water mixture is increased. Therefore, to maintain the same ratio of steam to water, it is often necessary to reduce the output. If the same output is maintained after a pressure reduction, the circulation rate will increase. No change in operating pressure should be made without consulting the boiler manufacturer.

Uneven Firing in Multiburner Boilers. In boilers using more than one burner, the chance of uneven firing exists. This, in turn, can affect circulation and possibly result in scale formation and overheating.

The burner service organization should be consulted to make any necessary changes.

The heavy formation of chips or flakes of scale is often caused by poor circulation. If the circulation is not improved, a chelate treating program should be considered.

Boiler Water Sampling. It is difficult to maintain proper boiler water conditions if the secured sample is not representative of the boiler water. Samples are frequently taken off the water-gauge column; this can result in condensate dilution.

The continuous blowdown line is an excellent sample source, but the sample line should first pass through a water cooler. If this is not done, a considerable amount of sample will flash into steam. The resulting water sample will indicate an incorrect high-TDS reading.

Extended Shutdown Periods. One simple change in operating procedure often aids in assuring cleaner boilers during long shutdown periods. Quite often boilers are drained while the boiler is hot. The boiler is then allowed to cool, after which the boiler is opened for inspection, etc. If a boiler has a high sludge potential, the sludge will bake on the surface while the boiler is cooling. A more effective method is to allow the boiler to cool to ambient temperature before draining. When the boiler is empty, it should be opened immediately and washed using a hose. With this procedure, the sludge does not bake on the surface.

Under some circumstances, where no pretreatment is used, scale may form in spite of proper internal treatment. When these conditions exist, it should be accepted that a change in polymer, etc., will not solve the problem. Instead, the boiler should be opened on a short, regular schedule (4–5 weeks) and washed to remove sludge. This, in turn, can prevent the development of scale (Mulloy, 1989).

Boiler Priming and Foaming

Both priming and foaming result in carryover of boiler water into the steam headers. Carryover is undesirable in that it causes erratic boiler operation, water hammer, deposits in superheaters, turbine deposits, etc. Also, the steam contaminant can interfere with film formation of filming amines.

Boiler Priming

Priming is caused by sudden, heavy steam demand; a swelling of the body of boiler water results. It can also occur at high boiler water levels; the high water level results in a reduced release surface. When priming takes place, the steam load should be reduced and the surface blowdown rate increased. Generally, priming cannot be controlled by changes in water treatment. The operating level of the boiler, of course, should be held at normal.

Boiler Foaming

Foaming is the formation of stable bubbles. High surface tension prevents the bubbles from breaking. Common causes of foaming are listed below.

High Alkalinity. Sodium hydroxide and sodium carbonate have a greater influence on foaming than neutral salts. When foaming occurs, an antifoam should be added or increased; also the continuous blowdown rate should be increased. The reason for the high alkalinity should be determined. It may result from ~~sufficient~~ blowdown; pretreated makeup water and ~~should~~ also be checked. Quite often, the source of ~~an o~~verdose of alkaline internal water treatment

~~the~~ steps covered under high alkalinity should ~~w~~hen high dissolved solids cause foaming. ~~of~~ antifoam are available but, for boiler use, the ~~two~~ types are polyamides and polyoxyalkylene gly- ~~pol~~yamides appear to perform best in low dissolved solid waters, while the polyoxyalkylene glycols are the choice in high dissolved solid waters.

High Suspended Solids. The most common cause of high suspended solids is high-hardness feedwater. Immediate corrective steps are to add or increase the dosage of antifoam compound. The continuous blowdown rate should be increased; the number of bottom blows should also be multiplied. Of the agents causing foaming, suspended solids probably have the least effect. Reasons for the increased hardness or other suspended solids should be determined. Refer to Part I, Pretreatment Performance Problems, and Chapter 8, Condensers.

Oil. Oil is a frequent cause of foaming. Check comment under Oil, pages 86–87.

Other Contaminants. Many other contaminants entering a boiler can cause foaming. If the condensate is contaminated, it should be dumped until corrective steps can be taken. As with other foaming problems, the continuous and bottom blow rates should be increased. Also antifoam should be added or increased.

6 Boiler Tubes

Boiler-Tube Failure

Tube failures in boilers involve either corrosion or mechanical failures, and at times both. Classification is made under what is considered the predominating element in the failure. In considering tube failures, it is important to note the location. For example: is the failure located close to the burners or soot blowers? Is the tube restrained by a boiler drum or tube support?

The largest number of water tube failures occur because of overheating. This accounts for about half of all failures. Fatigue or corrosion fatigue account for approximately one-quarter of the failures. Stress corrosion, pitting, corrosion, *hydrogen embrittlement*, etc. account for the remaining failures.

Steel tube and boilers generally depend on a *magnetite* (Fe_3O_4) film for protection from chemcial attack. Breakdown of this film is often involved in tube failures. Magnetite is soluble at a pH below 5.0 or above 13.0. Minimum corrosion takes place at a pH range of 9.0–11.0.

Failure Related to Internal Water Treatment

Chelants.   From their inception, chelants (Fig. 6-1) have been known to be corrosive if not used correctly. The two commonly used chelants are EDTA (sodium salt of ethylenediaminetetra-acetic acid) and NTA (sodium salt of nitrilotriacetic acid). A common error is to feed the chelants directly to the boiler drum. The chemical feed line entering the boiler drum will invariably corrode; both the concentration of chelant plus the elevated temperature in the boiler drum contribute to the aggressive attack. If the above chemical feed system is installed in an existing plant, it should be changed at the earliest possible moment. The chelant should be fed through a stainless steel line to the discharge side of the boiler feed pump.

Figure 6-1.   Chelant attack—cyclone separator.

Chelant attack, like acid attack, is characterized by clean dissolution of metal. Unlike chelant attack, however, the pits from acid corrosion are often undercut. Feedwater should be deaerated and treated with catalyzed sodium sulfite or hydrazine. When oxygen is present, it will decompose the chelant and aggravate corrosion.

Chelants should be used only where feedwater hardness is 2 ppm or less and well controlled. Where hardness control is erratic and few tests are made, it can be expected that corrosion will take place. It is not uncommon for severe attack to take place on drum baffles, steam separators, and steam-*generating tubes.* Free EDTA decomposes in the boiler drum; combined EDTA is stable. Free NTA decomposes to a lesser degree in the boiler drum. Tests for free chelant should be made on the feedwater instead of the boiler water. Normally, a residual of 1–3 ppm of free chelant should be maintained in the feedwater.

Optimum pH range for boiler water using EDTA is 9.0–11.5. Used correctly, EDTA provides a very adherent, protective magnetite film.

Chelants are effective in solubilizing ferrous iron. This accounts for the attack on metal surfaces. But chelants are not effective in controlling ferric iron under boiler pH conditions; they will not disperse or dissolve ferric iron oxides returning in condensate or entering with raw-water makeup. If appreciable amounts of iron oxide enter the boiler, it can be expected that deposit problems will develop; in high-heat-transfer areas, tube failures can result.

Whenever a change is made to a chelant program from a conventional phosphate treatment, the boiler should be reasonably clean. If not, the boiler should first be acidized; where this is not done, erratic chelant residuals can result. Corrosion can follow. Chelants for boiler scale control are a very effective tool when used correctly; they should not be employed in plants where poor controls exist.

Sodium Sulfite.   Sodium sulfite is used in the oxygen scavenging of the vast majority of boilers. Still, if the chemical is not used correctly, corrosion of boiler systems can result from oxygen and/or the decomposition of sodium sulfite.

The usual sodium sulfite residual in boiler water is 20–40 ppm. Sulfite should be added as far back in the system as possible; in most systems, this is the storage area of the deaerator. The fastest sulfite-oxygen reaction occurs at a pH of 9.0–10.0. Below and above these values, the reaction rates decrease. The reaction is also very slow at temperatures below 200°F.

If sulfite is added to the boiler drum, oxygen may react with the metal surface before it reacts with the sulfite. Sodium sulfite normally contains a catalyst to speed the oxygen-sulfite reaction; cobalt sulfate or chloride are most commonly used.

Where possible, the sulfite should be added alone to the deaerator storage. Many antifoam compounds, chelants, and polymers react with the cobalt catalyst, rendering it ineffective. Many water treatment problems involve a trade-off. Polymers and sodium sulfite are both more effective if added to the deaerator storage rather than to the boiler drum. The proper solution might be to evaluate the importance of the problem. If a corrosion problem exists, add the sodium sulfite alone to the deaerator and the polymer sludge conditioner to the boiler drum or feed line. On the other hand, if the main problem involves scale control, add the polymer to the deaerator storage area with the sodium sulfite.

At high boiler pressures, sodium sulfite decomposes; the usual decomposition boiler pressure limitation recommended is 900 psig. Under high-temperature/pressure conditions, generation of sulfides and sulfur dioxide can result in corrosion in both the condensate and boiler systems.

Also, there is evidence that decomposition takes place in the high-heat-transfer area of medium-pressure boilers. This can result in possible scale and corrosion in the boiler tubes. Since decomposition takes place only at the metal/scale interface, the condensate system is not affected.

Decomposition of sulfites in medium-pressure boilers most often involves oil-fired boilers; in the furnace area, oil-fired boilers release a much larger amount of radiant heat than gas-fired units. The solution to this problem is to change from sodium sulfite to hydrazine.

Sodium sulfite should not be relied on to protect a boiler from corrosion when the boiler is off-line for a weekend or overnight. The correct procedure is to maintain low steam pressure in the boiler or a *nitrogen blanket*. This prevents in-leakage of air.

When sodium sulfite is dissolved in a mix tank, only enough agitation should be used to dissolve the chemical completely. If agitation is continued, sulfite will react with air in the atmosphere.

When a sudden increase in sulfite dosage occurs, an investigation should be made to determine the source of increased oxygen in the system. Quite often, a malfunction of the deaerator is responsible. This is discussed in Chapter 4.

Caustic Soda. Where corrosion is concerned, caustic soda has some properties similar to sodium sulfite. Both chemicals assist in preventing corrosion in low- and medium-pressure boilers; both chemicals can cause corrosion in high-pressure systems.

Caustic soda is added as an integral part of a water treatment program. It can also be formed in a boiler as a result of the decomposition of bicarbonates or carbonates present in makeup water. As stated, caustic soda alkalinity (OH) in boiler water normally aids in the prevention of corrosion. Refer to Table 5-1, Guidelines for Water Quality in Industrial Water Treatment Boilers, for recommended caustic alkalinity. When caustic alkalinity is high, it can be reduced by increased blow-down. If low, it can be increased by the addition of caustic soda. Even at very high caustic alkalinities, few corrosion problems can be traced to the presence of caustic soda in low- and medium-pressure boilers.

In the period during which boiler drums were riveted, caustic soda concentrated under the rivets and caused *caustic cracking* (Fig. 6-2). This problem has virtually disappeared since very few riveted boilers are still in operation. Caustic cracking forms

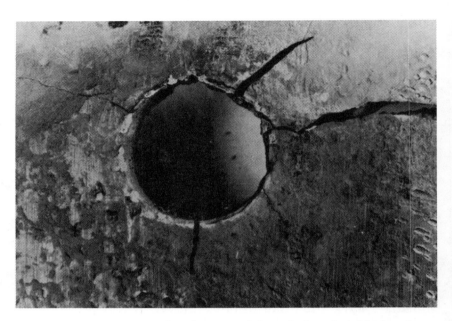

Figure 6-2.   Caustic embrittlement of threaded and welded carbon steel tube (courtesy NACE).

only in the presence of high alkalinities and stress. Still, it is possible to find cases of caustic cracking around rolled tubes. A small leak can cause buildup of alkalinity, while the rolling of the tubes induces the stress.

Chemically, caustic cracking can be prevented by the addition of 0.4 ppm of sodium nitrate per ppm of caustic soda alkalinity present in the boiler water. On the mechanical side, leaks should be stopped by rolling or welding.

The more common type of caustic corrosion is caustic gouging; this severe grooving or pitting attack is usually found in high-pressure boilers. Quite often, a buildup of black iron oxide is found in the vicinity of the caustic attack. Caustic gouging can also be found in the high-heat-transfer areas of medium-pressure water tube boilers. Any condition that favors the formation of high concentrations of caustic can result in caustic gouging. High heat transfer and poor circulation aggravate the condition. Attack generally takes place at pressures above 600 psig.

One type of caustic gouging occurs under what is known as departure from nucleate boiling (DNB). During nucleate boiling, steam bubbles form at distinct points on the metal surface. As the steam bubbles leave the surface, additional water washes the surface clean. However, as the steaming rate is further increased, the bubbles form faster than the surface can be washed. As a result, the concentration of caustic at that particular surface increases, and caustic gouging can result. As the steaming rate further increases, a stable *steam blanket* develops, and caustic gouging can occur along the edges of the blanket.

High concentrations of caustic soda are often found under porous deposits; high concentrations can also be found where steam blanketing occurs. This frequently takes place in nearly horizontal water tubes. One solution to caustic attack is to maintain clean surfaces. The water treatment program should be reviewed, and all possible steps taken to assure good circulation. Caustic soda, as known, should not be present in high-pressure boilers; in its place, the coordinated phosphate program should be used.

As previously mentioned, caustic alkalinity can be controlled through regulation of the surface blowdown. Alkalinity can also be reduced by various types of external treatment methods, including the hot-lime/soda process and ion-exchange methods.

In low-pressure boilers using calcium hard-water makeup, the carbonate-cycle type of internal treatment will also reduce alkalinity through the formation of stable calcium carbonate in the boiler.

Soluble carbonate salts such as sodium carbonate decompose in the boiler to form caustic soda and carbon dioxide.

Chromates/Nitrites. Chromates and nitrites are both oxidizing, anodic-type corrosion inhibitors. The widest use of both inhibitors is for cooling water systems. However, they are often used in low-pressure, fire tube heating boilers. Usually, very little makeup water is present when either chromates or nitrites are employed. Both chromate and nitrites inhibit corrosion by protecting the metal surface. Unlike a conventional sodium sulfite treatment, no effort is made to remove oxygen; since oxygen is not removed, the corrosiveness of the boiler water varies considerably, depending on such factors as salt content and pH.

When corrosion of tubes occurs during the use of chromate or nitrite, it indicates that the dosage is too low for that particular boiler water condition; the dosage should be increased until the corrosion protection is attained. With the use of either chromate or nitrite, a hydroxide excess of 200 ppm should be used. Sodium nitrite or sodium chromate residual should not be less than 400 ppm.

Chromates and nitrites should not be used in boilers operating above 200 psi, nor should they be used with organic sludge conditioners since they tend to oxidize these materials.

Coordinated Phosphate.   The coordinated phosphate program was designed to prevent the formation of excess hydroxide alkalinity in high-pressure boilers. Its greatest use is in the utility field. Where corrosion failures have occurred while this treatment is used, the reason has often been poor control. Often, these failures have taken place in utility plants operating under peaking conditions; these plants often operate with skeleton crews. Consequently, chemical control tests are often neglected. Under these conditions, caustic alkalinity develops, and caustic gauging or cracking occurs. The solution, of course, is to maintain the close control this method requires.

Other isolated reasons for failure have been reported (Frey, 1981, pp. 49–56). In some high-heat-transfer areas, it has been stated that large amounts of magnetite reacted with trisodium phosphate to produce free caustic. To prevent this type of attack, all steps possible should be taken to prevent deposit formation. The *coordinated phosphate* treatment is usually used in the 1000–1500-psi range (Peters, 1980). Above this pressure, a refinement known as the congruent phosphate treatment is used. While the coordinated phosphate treatment maintains a molar ration of less than 3.0 Na-PO$_4$, the congruent phosphate treatment operates at a ratio less than 2.6 Na-PO$_4$ (McCoy, 1981). As there is very little alkalinity present, serious upsets can take place if in-leakage occurs. Also, there is little alkalinity present to handle silica; it is imperative that silica levels be kept low. Where this is not done, silica scale can form, and possible tube failure can follow from overheating.

Acid-Cleaning.   Probably the largest amount of chloride-based corrosion comes from the incorrect use of hydrochloric (muriatic) acid for boiler deposit removal. Acid can be used for too long a period; 12 h is probably the maximum cleaning period. The acid-cleaning solution may be too hot; approximately 140°F should be the maximum temperature. Concentration of acid should be 15% or less. In addition, the effectiveness of the corrosion inhibitor should be determined. This can be done by submerging a steel nail in the acidified solution. If the nail corrodes with the evolution of hydrogen gas, the inhibitor is not longer effective.

Quite often, acid can be found in sludge or scale that has not been washed from the boiler. To assist in neutralizing this acid, a surfactant should be added to a solution of soda ash to assist in penetrating the deposit.

Boiler operators should attempt to acidize only low- or middle-pressure boilers using muriatic or sulfamic acid as the cleaning agent. Where the deposit is complex or the boiler pressure is high, a qualified contractor should be consulted.

Failure Related to Feedwater Quality

Oxygen.   The oxidation of steel is probably the best-known form of corrosion, but *oxygen attack* is not a common type of corrosion in operating boilers (Frey, 1981). Oxygen attack is characterized by pit formation covered with reddish nonprotective iron oxide deposits. The base of the pit usually contains black magnetic iron oxide, Fe_3O_4. Oxygen is also involved in the pitting attack, which often occurs during acid cleaning.

There are several reasons why oxygen is usually not involved in the corrosion of operating boilers. First, deaerators remove most of the oxygen in the preboiler cycle. When oxygen does enter a boiler drum, the boiler itself acts as a deaerator, and oxygen leaves the boiler with the steam. Finally, very few boilers operate without oxygen scavengers, such as sodium sulfite or hydrazine.

Nevertheless, boiler tubes or drums pitted from oxygen attack are not uncommon. As a solution, the first step should be to investigate the operating cycle of the boiler. It should be determined if the boiler is off the line during some portion of the day or on weekends. The resulting condensation of steam in the boiler drum creates a vacuum and draws in air; oxygen pitting can follow. Even maintaining a high residual of sodium sulfite will not prevent this type of attack. The most effective solution is to maintain a small amount of steam pressure or a nitrogen blanket during the period the boiler is off-line.

To repeat, only a small number of oxygen corrosion cases take place when a boiler is operating. To maintain this condition, the oxygen level in the feedwater should be kept at a low level. Field oxygen analyzers are now available through a number of industrial water treatment companies. The feedwater oxygen level should be checked at least once every year.

When the oxygen level is high (>8 ppb), an investigation should be made of the deaerator operation. In-leakage of oxygen at the feed pump is also possible. Refer to Chapter 4, Faulty Deaerator Performance.

Oxygen can also contribute to pitting corrosion during acid cleaning. High temperatures will aggravate the attack. Oxygen attack will also be increased if the acid corrosion inhibitor is adsorbed on scale present in the boiler. Under such conditions, the inhibitor will not be available to protect the metal.

To prevent oxygen corrosion in operating boilers, the oxygen scavengers should be added as far back as possible in the preboiler system. Usually, the location will be the deaerator storage area.

Sodium sulfite is usually catalyzed to speed the reaction with oxygen. To protect the catalyst, the sodium sulfite should not be mixed with other water treatment chemicals.

Bicarbonates and Carbonates.   Alkalinity in raw water is normally in the bicarbonate form. When the water is heated in a deaerator, the bicarbonate is converted to carbonate plus some caustic soda. Depending on boiler pressure, this reaction is carried further to completion in the boiler.

All the remarks pertaining to caustic soda that appeared earlier in this chapter under Caustic Soda also apply to bicarbonates and carbonates that are present in feedwater.

There is also a corrodant that develops in the decomposition of carbonates to caustic soda. Carbon dioxide is formed. This gas will be carried out of the boiler with steam to form carbonic acid in the condensate. The solution is to treat the condensate to neutralize the carbonic acid.

Chlorides.   Oxygen pits in boilers are aggravated by chloride ions. Chlorides are removed in demineralized makeup water for high-pressure boilers, but chlorides can still enter a high-pressure boiler system through leaks. For middle- or low-pressure boilers, no effort is made to remove the chlorides present in the feedwater. In fact, if a sodium cation ion exchanger is employed, the chloride level often increases through poor rinsing of the resin.

Every effort should be made to rinse the ion-exchange bed efficiently. When a demineralizer is used, the efficient use of this equipment will remove all chlorides.

The chloride content is especially important when a boiler is used in cycling operation; that is, the boiler may be shut down in the evenings or on weekends. Should oxygen enter the boiler, the chlorides present will aggravate the corrosion process.

Chloride content is also important when a chromate or nitrite boiler water treatment is used. When the chloride content rises, the amount of chromate or nitrite required for corrosion protection increases.

Organic Acids.   Organic acids are of importance only in high-pressure boilers where high-purity waters are used. Low- and middle-pressure boilers operate at too high an alkalinity for a small amount of organic acid to have any major detrimental effect.

Humic and fulvic acids are the most commonly encountered organic acids found in raw waters. They are formed from the decomposition of plants and animals.

Organic acids can lower the pH of high-pressure boiler waters (Davies, 1979). The solution is to take every step possible to prevent the entry of organic contaminants. New ion-exchange resins have been developed that are more effective than conventional resins in removing organic acids. Filtration and coagulation methods have also been used to remove organic contaminants.

Copper.   The role of copper in boiler corrosion is controversial. Many investigators believe copper is not active in the corrosion of an operating boiler.

When hydrochloric acid is used for acid-cleaning a boiler, copper can enter into the active corrosion of steel. The chemical reaction is as follows (Murphy, 1979):

$$2Fe^{+++} + 2Cu \rightarrow 2Cu^{+} + 2Fe^{++}$$
$$Fe + 2Cu^{+} \rightarrow 2Cu + Fe^{++}$$

The solution to this problem is to incorporate a complexing agent to combine with soluble copper. For this type of cleaning, it is recommended that a professional acid-cleaning company be contacted.

Even if copper or its compounds do not corrode an operating boiler, it serves no useful purpose. If nothing else, it acts as a heat-transfer barrier, or it can block the access of oxygen in differential oxygen corrosion cells. Most boiler water control charts limit the amount of copper allowed in feedwater.

Probably of greater importance is that the presence of copper in a boiler indicates that some piece of equipment in the pre-boiler cycle is corroding. Copper tests should be made on the effluent from equipment containing copper or a copper alloy. Soluble copper usually results from carbon dioxide, oxygen, or ammonia in the condensate system. The solution to contamination is to eliminate or neutralize the source of these gases.

Ammonia-Ammonium.   Ammonia is commonly found in high-pressure boiler systems as a result of the decomposition of added water treatment chemicals such as morpholine, cyclo-hexylamine, and hydrazine.

Ammonia also enters the feedwater as a result of leakage in a condenser. Depending on pH conditions, ammonia often enters the boiler as the ammonium salt.

In the boiler, the breakdown of ammonium ion results in the formation of ammonia, which leaves the boiler with the steam. The corresponding acid remains (Holmes and Mann, 1965). For example:

$$NH_4\,Cl \rightarrow NH_3 + HCl$$

Should the amount of contaminant be appreciable, the pH in the boiler will drop. Corrosion can follow. When lowering of pH occurs, alkalinity adjustments must be made to the boiler water until such time as the condenser leaks can be repaired.

Failure Related to Physical Conditions

Corrosion Fatigue.   Although *corrosion fatigue* is primarily an example of a mechanical tube failure, chemistry does play a role. Stresses caused by rapid thermal or pressure changes result in the initiation of cracks (Fig. 6-3). As cracks develop

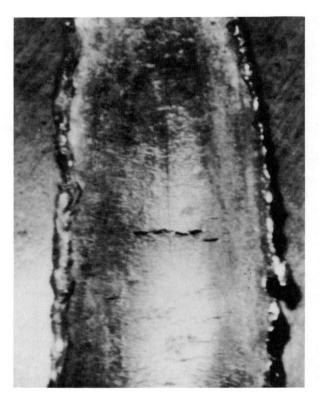

Figure 6-3.   Boiler corrosion fatigue.

from the flexing of metal, new areas are exposed to boiler water. The new oxide developed on the metal surface occupies a larger volume than the original metal. In the wedge, the oxides exert pressure and widen the cracks. In some systems in which the condition is not severe, it might take years before a failure occurs.

Cracks can develop around the circumference and also in a longitudinal direction.

Cracks around the circumference involve bending when thermal changes occur rapidly and usually when the tubes are restrained. Longitudinal cracks usually occur in the hotter boiler areas under rapid changes in pressure.

When corrosion fatigue cracks occur, it is usual to find other cracks parallel and close to the failure crack; the original wall thickness usually remains. The solution lies in trying to avoid peaking operations, load changes, or rapid startups with resulting shocks. Since the environment is involved, good water treatment practices should be maintained.

It is rare for a boiler tube to fail purely from fatigue. Most often, corrosion in one form or another is also present. Fatigue, unlike corrosion fatigue, is a purely mechanical type of failure resulting from stresses applied alternately. Since the environment is not involved, there is little that can be done in the way of water treatment. When pure fatigue takes place, there is usually only one main crack while, as mentioned, several parallel cracks are present when corrosion fatigue takes place. Mechanical preventive measures are the same as those recommended for corrosion fatigue.

Steam Blanketing.   Tube failure can result from *steam blanketing* for two reasons. Steam is a poor heat-transfer medium when compared to water or a water-steam mix. Accordingly, higher metal temperatures result in the high-heat areas. Also, as water evaporates to steam, the dissolved solids concentrate. Caustic and/or chelants and salts can then attack the metal. Blanketing can result from the separation of steam and water by gravity, centrifugal force, or irregular flow.

In straight, nearly horizontal boiler tubes, steam blankets can form on the upper part of the tube while water flows in the lower section. Corrosion can form in the crown of the tube through the concentration of salts. The same action can take place in the nearly horizontal sections that connect to the boiler drum. Blanketing can also result in the bends of steam-generating tubes through centrifugal action. The water is thrown against the outer wall of the bend. Steam on the inner side of the bend is exposed to the hot gases. Concentration of solids and overheating can result in failure in this area.

Boiler-tube welds can often cause steam blanketing (French, 1983, p. 61). Welds can upset the steam flow and result in the formation of a steam blanket; overheating of the tube can follow.

Regarding a solution to steam blanketing, there are limitations to modifications that can be made in water treatment. Since steam blanketing can result in high concentrations of salts, steps should be taken to maintain low alkalinity and/or chelate concentration. On the physical operating side, both heavy and very low loads should be avoided. Boiler design is very important in this type of failure but is beyond the scope of this book.

Overheating. Low-carbon steel is a solid solution of iron as the solvent and *pearlite* as the solute. Pearlite is a combination of iron and carbides of iron. Photomicrographs of normal boiler tubes indicate bands of pearlite precipitated in the iron. As the boiler tubes are heated between 900 and 1350°F, the pearlite becomes spheroidal. If heating of the tube is above 1350°F and reducing substances are present to react with the carbon, the pearlite is removed. The resulting metal is much weaker, and failure can result. Water treatment can play either a major or a minor role in overheating, depending on the particular cause of the overheating. Overheated boiler-tube failures are generally longitudinal (Barer and Peters, 1971). If a major blockage of water or steam flow occurs suddenly, the tube can fail in balloon fashion (Fig. 6-4); thin-edged failure results in high-heat areas. Scale breaking loose is one cause of tube blockage.

Sudden overheating can also result from the temporary blockage of tube circulation during too rapid startup. Increased *firing rates* can also interfere with circulation to the point at which sudden rupture of tubes occurs. Low water level in the steam drum can also interfere with circulation and cause overheating failures.

Caustic gouging and acid can also cause thin tube walls. Consequently, knife-edged failures (Fig. 6-4) are not exclusively a result of sudden overheating. Nevertheless, sudden overheating remains the principal cause.

Gradual scale increase can result in *creep* failure from over-

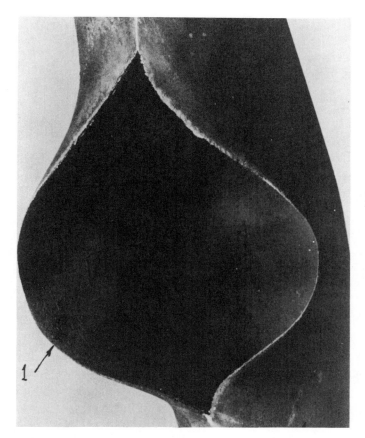

Figure 6-4.   Burst boiler tube—thin-edge rupture (courtesy of Gordon and Breach Science Publishers, *Why Metals Fail,* Barer and Peters, 1970).

heating (Klein, 1973). Under these conditions, the tube failure is longitudinal, but the edges are thick-walled (Fig. 6-5) instead of knife-edged. Often, a tube will start to bulge, and the scale in that area will crack and wash away. The subsequent cooling stops further expansion of the tube. When a tube is thought to have cracked from creep, other tubes in the area should be inspected for signs of bulging. If bulging is found, steps should be taken to remove the scale in the boiler.

Figure 6-5.   Burst boiler tube—thick-walled rupture (courtesy of Gordon and Breach Science Publishers, *Why Metals Fail*, Barer and Peters, 1970).

Finally, it should be noted that, when a change is made from gas firing to oil, the amount of radiant heat increases; chances of overheating developing are increased.

High-Heat-Transfer-Hideout.   In high-pressure boilers, it is occasionally noted that, as loads are reduced, the concentration of certain salts will increase. This can be attributed to a condition known as "hideout." In certain localized areas at high loads, the high-heat transfer results in the precipitation of salts that are normally soluble. As the load is reduced, the salts that have precipitated are washed back into the solution. One undesirable result might be an increase in silica higher than recommended levels. During the period of high load, there is also the danger of overheating, hydrogen embrittlement, and caustic gouging.

From a water treatment viewpoint, the only remedy is to maintain feedwater quality to the point at which hideout is reduced or eliminated. If firing rates can be reduced, of course, that will also prevent hideout from occurring.

Excessive Steam Load-Circulation.   When the steam load on a boiler is increased beyond its designed capacity, circulation can be affected. The number of tubes acting as *risers* can increase to the point at which sufficient water is not supplied to the downcomers. Stagnant circulation can result, with the possible development of overheating. The obvious solution is a reduction in steam load.

Failure Involving Hydrogen Embrittlement

Hydrogen can originate from several different chemical reactions in a boiler. Physical factors are also involved in its generation. For these reasons, plus its importance in boiler-tube failure, hydrogen is listed separately from other types of failure.

In cases of hydrogen embrittlement, heat input and circulation are of great importance. Hydrogen embrittlement takes place only under high-heat-input conditions. It is most often found in boilers operating above 500 psi although it is occasionally present in the high-heat-transfer areas of lower-pressure boilers. Hydrogen gas formation often occurs under hard, dense deposits where high temperatures and high-chemical-concentration conditions exist. Atomic hydrogen enters the metal. In voids, it reacts with carbon in the steel to form methane gas. The decarbonization weakens the steel while the trapped methane gas exerts pressure.

Underneath deposits at high temperatures, iron in steel can react directly with water in the form of steam.

$$3Fe + 4 H_2O \rightarrow Fe_3O_4 + 4 H_2$$

Concentrated caustic can also enter into the production of hydrogen in the following manner. It first reacts with the protective magnetite film (French, 1983, p. 189):

$$4 NaOH + Fe_3O_4 \rightarrow 2 NaFeO_2 + Na_2FeO_2 + 2H_2O$$

With the protective magnetite removed, the caustic then reacts with iron as follows:

$$Fe + 2 NaOH \rightarrow Na_2FeO_2 + 2H$$

Hydrogen can also form in the reaction of concentrated free chelant with steel. Occasionally, acid remaining from acid cleaning can attack the boiler tubes to produce hydrogen.

Hydrogen embrittlement of boiler tubes is often characterized by a detachment of a portion of the tube (ASME, 1972). This type of failure is often called a "window" fracture. Further identification of hydrogen embrittlement should be done by a qualified metallurgical laboratory.

One preliminary screening test for hydrogen embrittlement can be conducted on site. This test consists of placing a small boiler-tube ring being tested in a vise, with the fired section of the ring facing the vise. As pressure is exerted by the vise, the inside of the tube will be under tensile stress. If hydrogen embrittlement is present, the inside of the tube will readily crack.

Unlike sudden and creep overheating, little ductile deformation is present in the case of hydrogen embrittlement. The "fish-mouth" appearance of tubes failing from overheating is absent in the case of hydrogen embrittlement.

Severity of hydrogen embrittlement failure varies considerably. Some failures occur in very short periods of time while, in mild cases, many years may pass before failure occurs. The solution to hydrogen damage is to maintain clean surfaces. Scale and other factors that contribute to overheating should be avoided; this includes poor circulation.

Corrosion During Shutdown Periods

The vast majority of boiler corrosion problems take place when the boilers are off-line for either short or extended periods. If a boiler is off-line for a weekend or overnight, a small amount of steam pressure should be maintained to prevent the entrance of air. As an alternative, a nitrogen blanket should be maintained. When nitrogen gas is used, extreme care must be taken before anyone enters a boiler drum for an internal inspection. Although nitrogen is not a poisonous gas, it is still dangerous since it displaces oxygen. The gas supplier or a safety equipment company can usually supply safety procedures regarding its use.

Although it is often used, a high sodium sulfite residual is not recommended for standby protection if the boiler water remains at operating level. As a boiler cools, it creates a vacuum, and air enters. Oxygen will react at least with sulfite at the air-liquid interface. Pitting at the water line often occurs. Also, when operation is resumed, the oxygen-saturated water will be pulled into the downcomer tubes.

For extended idle periods, the boiler can be stored dry or wet. When dry storage is used, a nitrogen blanket can be maintained or quick-lime trays can be placed in the boiler and the manholes sealed. Hydrazine has often been used successfully in wet storage to passivate the metal. The procedure is to add 1 gal of hydrazine (35%), plus 3 lb of caustic soda per

1000 gal of boiler water. Heat for 3 h at 180°F. This treatment provides an excellent protective film. The boiler should be drained before it is placed on-line. If this is not done, the heavy hydrazine concentration will break down to form ammonia at boiler operating pressure.

Boiler Superheater Tube Failure

Superheaters are not as forgiving of operating engineering errors as boiler tubes are. High temperatures are involved, and heat transfer to saturated steam is not nearly as rapid as to boiler water. Operator error, such as too rapid startup, can sometimes result in little or no flow of saturated steam through the superheater tubes; overheating then occurs.

Selection of metal alloy, superheater location, etc. is not the responsibility of the operating engineer. But, in view of such conditions, it is essential that errors not be committed in water treatment operations that will adversely affect the superheater.

Oxygen Pitting.   Oxygen pitting is found primarily in cycling boilers. When a boiler is taken off-line, a small amount of steam pressure must be kept on the unit, or a nitrogen blanket must be maintained. If these precautions are not taken, steam will condense in the superheater tubes; the water plus oxygen will result in corrosion of the superheater tubes during the idle period. When oxygen corrosion takes place, pits will form. At the same time, metallurgical examination will indicate normal structure—ruling out overheating.

Specifications of new boilers should call for drainable superheaters.

Creep.   Failure from creep (Fig. 6-5) usually occurs over a long period of time, during which metal is heated slightly over its design temperature and pressure. In most cases of superheater creep failure, deposits on the ID surface contribute to failure; thick-lipped edges are usually present when creep failures occur. In some creep cases involving superheaters, the bulging results in a larger-diameter tube. Steam velocity is decreased, and rapid overheating and failure can follow. A thin, knife-edged, longitudinal split can then result (Fig. 6-4). All water treatment practices that aid in maintaining clean surfaces should be followed.

Carryover.   Two main factors account for overheating failures in superheaters. Steam on the ID side resists the transfer of heat to a much greater degree than water. As a result, the superheater metal is subjected to high temperatures. High temperatures also result from deposits on the ID side.

Every step possible should be taken to avoid carryover. Refer to comments in Chapter 5, under the heading Boiler Priming and Foaming. Preventive measures include avoiding heavy loads or rapid changes in load.

Flame Impingement.   Faulty burner adjustment or too short a distance between the burner and the radiant superheater tubes causes flame impingement (Shields, 1961, p. 252).

As with many other overheating failures, any deposit or obstruction in the superheater tubes will aggravate the problem. Failed tubes will be balloon-shaped and thin-lipped. Solutions to the problem are essentially mechanical: included are adjusting the burner, angling the burner, and increasing the distance between the burner and radiant superheater.

Caustic Stress Cracking.   Caustic embrittlement results from carryover of caustic from the boiler drum. Concentration of caustic in the superheater tubes in the presence of stress results in the formation of caustic stress cracking. The cracks are *intergranular* and usually longitudinal. Heavy loads and all other factors that cause carryover should be avoided. Boiler water alkalinity should be kept low.

Short-Term, High-Temperature Tensile Stress Failures.   As mentioned previously under Creep (page 118), a failure often starts slowly as creep and then accelerates as temperatures increase. Tube blockage, flame impingement, etc., can also result in short-term failure. The tube will expand in balloon fashion, and thin-edged failure will result. As in other failure cases, all steps to prevent ID deposits should be taken.

Tube Bend Failure.   Scale flakes often collect at bends in the superheater tubes. Rapid longitudinal thin-edge failure can occur from overheating of the metal. The solution lies in the prevention of scale. The quality of the steam entering the superheater should be kept as high as possible.

Start-up.   Too fast a start-up can result in rapid burnout of radiant superheater tubes. During start-up, when little or no steam is flowing through the superheater, the gas temperature entering the superheater should be kept at 900°F or lower (Trigs, 1978). The manufacturer's instructions on start-up should be followed. All steps to avoid deposits in the tubes should be taken.

7 Turbines

Turbine Deposits

Silica. A major component of deposits found in steam turbines is silica. The solubility of silica in steam decreases as the steam is supersaturated; vaporization occurs mostly at pressures above 600 psi (Shields, 1961, p. 229). As would be expected, the solubility of silica decreases in the colder sections of the turbines—below 520°F. The deposit is usually gray-white and hard. Buildup of silica results in a decrease in turbine operating efficiency.

The initial step in preventing silica deposits is to control the silica concentration in the boiler water. Refer to Table 5-1 for water quality guidelines. This table lists the recommended boiler water silica limits at various boiler pressures. In addition, it should be kept in mind that increasing the boiler water pH will decrease the amount of silica vaporizing in the steam (Sohre, 1972).

Silica in the superheated steam should be controlled at 0.03

ppm or less. Maintaining a high pH is a limited method of controlling silica in the steam since pH of high-pressure boilers has to be controlled to prevent caustic attack. Refer to Coordinated Phosphate, page 104.

Silica in steam can result from both mechanical carryover of boiler water and vaporization of silica. Any silica present in mechanical carryover will vaporize in the superheater up to the solubility limit at that operating pressure. At 1000–1500 psi, all the silica deposits found on turbine blades will be in the form of amorphous silica and will not be water-soluble.

The most effective solution to silica deposit problems is prevention. When silica limits in the boiler water are exceeded, the reason for the excess should be determined. Since *amorphous* silica is water-insoluble, it is usually removed by blasting with aluminum oxide grit.

Salts—Caustic—Metal Oxides.   Silica in its various forms probably makes the greatest contribution to turbine deposits. But sodium chloride, sodium sulfate, sodium hydroxide, and various other salts are also present in most turbine deposits. Their major contributions to turbine deposits are made in industrial units operating at 600 psi or less. Iron oxides and copper oxides are also present, probably to a greater degree in high-pressure utility boilers than in industrial turbines operating at lower pressures.

Most deposits occur in the intermediate turbine stages. At this point, temperatures have started to drop while, at the lower pressure end, moisture has started to wash away soluble deposits.

The *steam quality* from industrial boilers is lower than from high-pressure utility boilers; accordingly, higher caustic and salt deposits can be expected to form on turbine blades in industrial plants. Caustic deposits can form over a broad range of turbine operating temperatures and pressures, while sodium chloride and sodium sulfate form over a narrower range. Water-soluble deposits can best be prevented by controlling steam purity.

At boiler pressures of 1500 psi and below, the contaminants of steam originate as carryover in the steam moisture. As steam pressure increases, the properties approach water, and the solubility of the salts and metal oxides increases in the steam. When the steam passes through the turbine, the pressure drops and the salts and oxides precipitate.

In high-pressure units, it is often necessary to maintain total dissolved solids below 0.03 ppm in the steam to prevent water-soluble deposits from forming. When deposits have formed, the most common method of removing them is to reduce to lower load and slow speed and to introduce condensate to the steam; the *wet steam* will dissolve soluble salts. Operating on the same principle, turbines that shut down frequently do not often experience soluble-salt deposits. The development of wet steam at shutdown washes the blades.

Another effective method of cleaning turbines in industrial units in the 400–700 psi range is to introduce morpholine at an increased dosage. The morpholine appears to act as a surfactant and prevents caustic and soluble salts from adhering to the blades (Mitchell and Schroeder, 1964). It also aids in removing deposits.

Iron and copper oxides form from corrosion in the steam preboiler cycle. High-pressure feedwater heaters are the usual source. Control involves all steps that reduce iron and copper levels.

Changes in operating conditions probably contribute to the greatest amount of deposit carryover. These changes are brought about by load changes, start-ups, oxygen in-leakage, boiler blowdown, etc.

Turbine Corrosion

Most of the elements necessary for corrosion to occur are present in turbine operation. These include high temperatures, phase changes, thermal changes, vibration, and suspended abrasive particles. Accordingly, many of the various types of corrosion are present. These include pitting, *stress corrosion cracking, erosion-corrosion,* and corrosion fatigue.

Most of the corrodants in a turbine originate upstream from the turbine. Accordingly, maintaining excellent water treatment conditioning in the boiler and preboiler cycle is very important. Oxygen can originate both upstream from the turbine and in the turbine. Water droplets, which can cause severe *erosion,* originate in the turbine.

Design and materials of construction are of prime importance in the control of turbine corrosion, but the scope of this book is limited to water treatment. A major step in the prevention of turbine corrosion is to keep the steam as free of contaminants as possible. Impurities can result from carryover, condensate leakage, oxygen in-leakage, etc.

Caustic Stress Cracking.   Caustic is probably responsible for the largest amount of turbine corrosion; its presence can result in stress corrosion cracking. Caustic is found throughout a wide range of the turbine cycle. Only in the low-pressure wet area is it dilute enough to cause no damage.

The source of caustic in high-pressure boilers is usually misoperation of the coordinated phosphate treatment or leakage from a demineralizer. Condenser leakage can also result in increased caustic in the steam. In medium-pressure boilers, all steps necessary to prevent boiler water carryover should be taken. Refer to Chapter 5.

Chlorides (Stress Corrosion Cracking, Corrosion Fatigue, Pitting).   Chlorides, in addition to caustic, are an important cause of stress corrosion cracking. However, in turbines, the range of chloride cracking is very narrow compared to caustic. Chlorides are usually found only in the wet/dry transition zone (Wilson line). Cases of chloride-related cracking have also been found in the crossover bellows.

In addition to the location, an analysis of a deposit in the area can assist in determining the cause of cracking. Finally, a metallurgical analysis can assist in differentiating between caustic and chloride cracking.

Chlorides and other salts are involved in corrosion fatigue problems. Any areas subject to flexing tensile stresses are subject to corrosion fatigue attack. Pitting attack from chlorides frequently occurs in the blade or nozzle surfaces. Problems often develop during layups, when oxygen can leak into the turbine casing. The relative humidity of the air during this period is also an important factor in the rate of pitting. Relative humidities above 60% should be avoided. Dry air should be present if possible.

Chlorides can enter a high-pressure system through condenser leakage or when a demineralizer operates past its regeneration point. At lower boiler pressures, failure to rinse an ion-exchanger softener sufficiently can result in an increase in chlorides in the boiler water. Total dissolved solids, including chlorides, in the boiler water may also be higher than recommended for the prevailing pressures. Any condition contributing to priming or foaming will contaminate the steam. Refer to Chapter 5.

Prevention of chloride corrosion problems involves taking all steps necessary to maintain pure steam. If carryover does take place, phosphate boiler treatments tend to reduce chloride-related corrosion. Where phosphate salts are not present, the corrosion rates tend to be higher.

Sulfides. The decomposition of sodium sulfite in high-pressure boilers results in the formation of hydrogen sulfide and sulfur dioxide. The end result is cracking of stressed low-alloy steels. Rotor dovetail cracking has been reported from this cause (Lindinger and Curran, 1982).

Sulfide stress cracking is not a major problem at present since hydrazine has replaced sodium sulfite in the vast majority of boilers operating over 600 psi.

Erosion-Corrosion. Erosion-corrosion results from solid particles or water droplets in the steam removing protective metallic films, which exposes the metal to corrosive attack. Silica is often the erosive medium in high-pressure steam. It forms from physical carryover of boiler water. In superheaters, all moisture is removed; part of the silica present is dissolved in superheated steam, while the remaining silica is carried as suspended particles in the steam. The action of the suspended particles is similar to sandblasting.

The solution lies in maintaining silica in the recommended range for the operating pressure. One common source of silica is surface raw water, which contains colloidal silica in addition to ionized silica. The colloidal silica is not removed in the demineralizer but, in the boiler, it is converted to the ionized form.

Erosion—Suspended Particles—Water. Erosion (Fig. 7-1) is similar to erosion-corrosion except that no corrosion takes place. The damage is purely mechanical. Like corrosion, the end result is the destruction of metal. In some cases, such as silica, it can be controlled through water chemistry. Any controls that reduce the suspended solids will assist in reducing erosion from that source. One other type of erosion experienced in turbines is caused by water droplets found in the low-pressure areas of turbines. Damage tends to resemble the honeycomb appearance of *cavitation.*

Carryover of boiler water to a turbine during start-up can frequently be detected by vibration accompanied by rumbling.

Erosion from water droplets (Fig. 7-2) will, under normal conditions, form only in the low-pressure section of a turbine, while erosion from solid particles is found in the high-pressure section. Damage from water droplets is usually found on the moving blades rather than the stationary blades. Unlike solid-particle erosion, water chemistry has no effect on water-droplet erosion.

Oxygen. Oxygen in-leakage can contribute to pitting. The source of the leakage should be determined and corrected.

Figure 7-1. Solid-particle erosion—turbine blades (reprinted with permission from *Power*, Sept. 1976, copyright McGraw-Hill, Inc., 1976).

Humic-Organic Acids. If no preventive measures are taken, humic acid and other organic acids from surface waters are capable of contributing to attack on turbines. On the other hand, virtually all systems operating turbines employ neutralizing amines, such as morpholine or cyclohexylamine; as a result, the organic acids are neutralized.

Frequently, seasonal changes account for increases or decreases in humic acid present in the surface water. Consequently, the neutralizing amine demand varies.

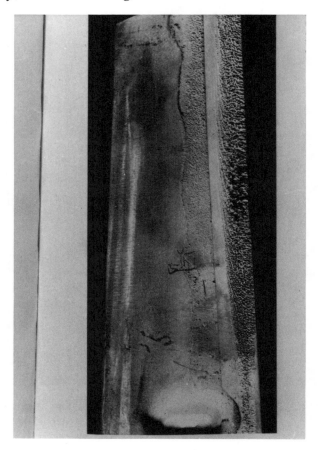

Figure 7-2.   Water drop impingement erosion of steam turbine blade (courtesy of NACE, Houston).

8 Condensers

Steam-Side Condenser Corrosion

Unlike most components in the steam-water cycle, there are no major deposit problems on the steam side of a steam condenser. But corrosion problems are very much in evidence.

Impingement. This corrosion mechanism (Fig. 8-1) is primarily the mechanical erosion of soft metals although chemical corrosion also plays a role. In condenser operation, only aluminum and copper-bearing metals are affected (Syrett and Colts, 1943). High-moisture steam and high velocities, most often found in the peripheral areas, are primarily responsible for the damage. In the early part of the attack, the surface metal is highly polished but becomes rougher as attack continues. Eventually, the tubes can fail. Unfortunately, there is no water treatment solution to this type of corrosion. Stainless steel and titanium tubes are frequently used in condensers in place of the softer metals. They do not seem to be affected by impingement

Figure 8-1.   Copper impingement.

in condenser operation. Adversely, they are more expensive and do not transfer heat as readily.

One answer to this steam-water corrosion problem lies in installing baffles. Stainless steel clips have also been placed on tubes to protect them against impingement.

Corrosion Fatigue.   In this type of attack (Fig 8), mechanical vibration of the tubes is caused by the flow of steam from the turbine exhaust into the condenser. Chemical attack plays only a minor role. Cracks in the tubes eventually develop. Solutions to the problem are mechanical. Proper spacing of the tube support plates and proper baffling of tubes, etc., are involved.

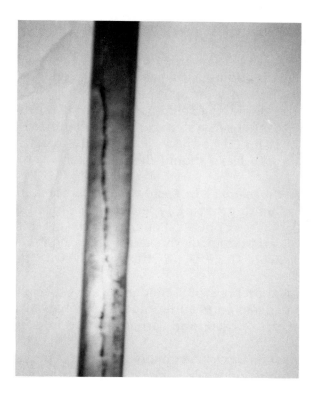

Figure 8-2.   Brass stress corrosion cracking.

Titanium and copper-bearing alloys are more susceptible to
ιthis type of attack than stainless steels.

Stress Corrosion Cracking.   This type of attack (Fig. 8-2) is
caused by tensile stress plus a specific environment (Wilson,
1961). On the steam side of condensers, the specific environment is usually ammonia or various types of amines. In condenser operation, only copper alloys appear to be affected on
the steam side. It is impossible to avoid ammonia in high-
pressure systems because the amines used for condensate pH
control all decompose to some degree. Ammonia alone is also
used for condensate pH control; hydrazine, which is used for

oxygen scavenging, also decomposes to ammonia. Oxygen increases the ammonia attack.

All steps possible to reduce ammonia and oxygen should be taken. Chemical additions that contribute to ammonia production should be held to a minimum.

It would be desirable to eliminate oxygen. Unfortunately, all leaks into a condenser system cannot be eliminated. Tests for oxygen leaks should be made on a regular basis. Corrective steps to stop the leaks should then be taken.

Condensate Corrosion.   In areas where ammonia and oxygen can concentrate, a grooving type of corrosion can occur (Wilson, 1961). Only copper-containing alloys are affected; the presence of carbon dioxide can cause grooving in both copper alloys and steel. There is no stress involved in this type of attack.

Air-removal sections of a condenser are especially susceptible to condensate corrosion. Solution involves all steps that hold ammonia, oxygen, and carbon dioxide to a minimum.

Noncondensable Gases.   Ammonia, oxygen, carbon dioxide, and other noncondensable gases can contribute directly to corrosion in a condenser (Bernard, 1981). Also, the gases, including nitrogen, interfere with heat transfer and exert back pressure on the turbine, decreasing its efficiency.

Air has a very high resistance to heat transfer. It is generally recognized that a layer of air only 0.04 in. thick gives the same resistance to the flow of heat as a 1-in. thickness of water.

Ammonia can cause stress cracking and general or grooving corrosion on copper alloys. Ammonia tends to concentrate in the lower and cooler section of a condenser. When loads are lower than normal, even cooler conditions exist in the lower sections, and more ammonia condenses. No quick solutions are available, except to make certain that as much ammonia as possible is being removed in the deaerator and that excess dosages of neutralizing amines or hydrazine are avoided.

On a long-term basis, a change in the type of tubing used may have to be considered or the steam-flow patterns changed.

Carbon dioxide contributes to general or grooving corrosion on both steel and copper alloys. Oxygen causes pitting and also contributes to the stress corrosion cracking of brass.

These gases are eliminated by air-removal equipment, such as steam-jet ejectors and mechanical pumps; the air-removal outlet is located in the lower part of the condenser. Discarding after cooler drains rather than returning them to the condenser also aids in reducing noncondensable gases to a minimum (O'Keefe, 1973).

Noncondensable gases in the condenser originate from leaks into the condenser, release of air from water, and water treatment chemicals. Likely areas of air leaks are glands, stuffing boxes, flanges, etc.

Testing for leaks is usually done by spraying a test gas in the area of the suspected leak. Downstream, a test can then be made for the tracer gas. Helium has been used for this purpose. A candle flame can also be used to detect leaks while a system is under vacuum.

A considerable amount of air leakage can be detected by shutting off the air ejector or pump for about 10 min. If a large amount of air leakage exists, the condenser pressure will increase rapidly.

Condensate System Contaminants

Condensate water is very low in mineral content and, in most cases, is low in oxygen content. These properties make it very desirable as a boiler feedwater. This does not mean that it is always desirable to recover condensate. Pumping or piping costs may not make its recovery economical. Chances of contamination may also rule out the recovery of condensate. When condensate is recovered, some problems still exist. The presence of carbon dioxide, oxygen, and ammonia can have adverse effects on metallic materials of construction. This corrosion, in turn, can result in the return of undesirable contaminants to the boiler.

Carbon Dioxide.   The primary source of carbon dioxide is the decomposition of bicarbonates and carbonates in boiler water. With the formation of condensate, the gas dissolves to form carbonic acid. Since condensate has very little buffering capacity, the pH drops into the acidic range. Attack on iron- and copper-bearing metals follows. Carbonic acid attack is characterized by general thinning of the metal. The threads of piping are usually affected first; this area is the thinnest section of the piping system, and the threads are also under stress.

Methods of preventing attack are several. Since carbon dioxide originates in many cases for bicarbonates in raw water, the source of makeup water should be checked; a lower alkalinity water might be available.

Pretreatment methods for lowering bicarbonate alkalinity include hot- or cold-lime treatment or several types of ion-exchange dealkalizers. Where a low-pressure boiler (<200 psi) is in use, the carbonate cycle internal treatment method will reduce the carbon dioxide potential if a significant amount of calcium hardness feedwater is involved in the operation. If the feedwater is softened, the reduction in carbon dioxide will be insignificant.

With existing equipment, a number of mechanical methods can be considered to hold carbon dioxide corrosion to a minimum. Heading this list is the operation of the deaerator.

The best method of testing the deaerator operation is to make an oxygen determination on the effluent. If oxygen is removed efficiently, it can be assumed that free carbon dioxide will also be removed. Carbon dioxde present as carbonate or bicarbonate will not be removed. Through the elevated temperature, there will be some decomposition of carbonates and bicarbonates to free carbon dioxide, but the greatest reversion will take place in the boiler.

If the deaerator has an external vent condenser, the drains should be discarded—not returned to the deaerator storage area; the drains are usually saturated with carbon dioxide and oxygen.

Traps in the condensate return system should be checked to make certain they pass carbon dioxide gas as well as condensate (Monroe, 1983). As far as corrosion is concerned, traps should not be used that subcool the condensate. Traps of that type will dissolve more carbon dioxide. A buildup of red iron oxide in a trap is one indication that corrosion is taking place in the condensate system. In addition to traps, uninsulated condensate lines will also contribute to subcooling.

If a steam condenser is installed in the boiler system, the air-removal equipment performance should be checked. When all mechanical steps have been taken and the pH of the condensate is still low, then chemical treatment steps will have to be taken. Usually, the volatile amines, such as morpholine or cyclohexylamine, are employed; the choice depends on the desired distribution ratio of the amines. This is the ratio of the amine solubility in the steam to the solubility in water. Morpholine has a low ratio, and so it tends to be dissolved in the first condensate formed. More morpholine than other generally used neutralizing amines is lost in boiler blowdown. To avoid this loss, it is often fed to the steam header instead of the feedwater line. In utilities, morpholine is typically used to protect turbines. Cyclohexylamine has a higher distribution ratio, so that it dissolves later in the system. Cyclohexylamine is affected by pressure to a greater degree than other amines.

Where high concentrations of carbon dioxide are present, the use of cyclohexylamine often results in the formation of white deposits. If this occurs, a change to a different type of amine is recommended. Combinations of morpholine and cyclohexylamine are often used. *Diethylaminoethanol* and *aminomethylpropanol* are two others types of amines employed. Aminomethylpropanol should not be used in boilers operating at 100 psi or less. At these low pressures, the amine will not distill from the boiler. Using neutralizing amines, a desirable pH range of the condensate is 8.2–8.5.

Measuring methods used to determine pH are vitally important in condensate systems. Since condensate contains virtually no buffering agents, it is very sensitive to any indicator added during the pH determination. The least accurate method of measuring pH in condensate is pH paper. It is followed by pH indicator solution. If at all possible, a pH meter should be used to determine condensate pH. This is the most accurate measuring method if the electrodes and meter are well maintained. A calomel electrode should be used in determining condensate pH instead of the silver-silver chloride electrode. Combination calomel electrodes are commercially available.

Equally important is the sampling method. If a condensate sample is secured and taken to a testing location, the pH can change through the addition of carbon dioxide from the air or the release of any ammonia present in the condensate. A more representative sample is using an open beaker; overflow is directed to a drain. The pH electrodes are placed at the bottom of the beaker.

As mentioned, the pH of the condensate should be maintained between 8.2 and 8.5. A neutral pH of 7.0 does not assure that carbonic acid attack is not occurring. The action of carbonic acid corrosion of steel will raise the pH close to 7.0. A pH of 8.2–8.5 will assure that no carbon dioxide attack occurs in the sampling area.

Measurements should be taken at several points in the system. If different results are obtained, it may be necessary to use a blended amine or to add amines at several points in the system. Where high amounts of carbon dioxide are present, the use of a filming amine should be considered; the most widely used filming amine is octadecylamine. One commonly used salt is octadecylamine acetate. Other emulsion forms are available. Combining a filming amine with a neutralizing amine improves performance of the filming amine.

Filming amines should be mixed only in hot condensate; they should not be mixed with other chemicals or raw water. Filming amines have a tendency to remove any metallic oxide deposits in a newly treated system. Plugged lines and traps can result. In addition, filming amines tend to degrade in the presence of ferric iron. For this reason, the product should first be fed at a low dosage and gradually increased as the system is cleaned. Carryover of boiler water will interfere with the performance of the filming amines.

Oil and/or iron oxides plus filming amines can result in the formation of sticky deposits in both the condensate lines and the colder sections of a boiler (Trace, 1980).

When condensate systems are shut down for extended periods, such as the summer season, consideration should be given to the use of a nitrogen blanket if that is possible. Filling a system with a solution of alkaline chemical such as soda ash can assist in reducing corrosion. If some protective step is not taken, the combination of moisture and air will result in corrosion during the idle period.

Oxygen.   On a molar basis, oxygen is about 8–10 times as corrosive as carbon dioxide in condensate. Unlike carbon dioxide, corrosion from oxygen is characterized by pits rather than general corrosion. A mixture of oxygen and carbon dioxide is even more corrosive than oxygen alone.

All the mechanical methods of removing carbon dioxide also apply to oxygen.

Neutralizing amines only indirectly inhibit oxygen attack in condensate lines. Where both carbon dioxide and oxygen are present, the neutralizing amine will remove the carbon dioxide, making the oxygen less corrosive.

The best chemical method of keeping oxygen out of the condensate system is to remove it in the preboiler system. Sodium sulfite and hydrazine are the most commonly used chemicals to remove oxygen. Filming amines do not remove oxygen from condensate, but they do provide a protective film for the metal (Gelosa and McCarthy, 1979). Where pits are already present in a condensate system, the filming amines are not very effective in protecting that area from further corrosion. Dosages for both neutralizing and filming amines are difficult to predict since they both recycle through the deaerator. The reacted neutralizing amine decomposes in the deaerator to form carbon dioxide plus the amine. The gas is lost through the deaerator, while much of the amine recycles.

Ammonia (Including Neutralizing Amines and Hydra-zine).   Most of the ammonia present in condensate is the result of the addition of ammonia or the decomposition of neutralizing amines to form ammonia. Decomposition of hydrazine and organic contaminants also contributes to the formation of ammonia. In a boiler system, ammonia is normally corrosive to copper-bearing metals.

All mechanical methods of reducing carbon dioxide (see pages 134–137) also apply to ammonia. Chemically, the only control is to maintain ammonia, neutralizing amines, or hydrazine at minimum levels. These products should not be overfed or slug-fed.

Hydrazine is virtually always used in high-pressure boilers (>600 psi) since sodium sulfite decomposes. It is also often used in lieu of sodium sulfite in many medium-pressure boilers. The usual hydrazine residual in boiler water is 0.01–0.20 ppm. If the residual is in excess of 2 ppm, it is likely to decompose to form ammonia. This enters the steam condensate system and can react with copper or copper alloys. Maintaining the correct boiler residual corrects this problem. Hydrazine is a known carcinogen, and so steps should be taken to keep vapors from the surrounding air.

Oil.   In boilers, oil in condensate can contribute to scaling. In combination with filming amines, it can form sticky deposits in both the condensate system and the boiler.

Condensate containing oil on a continuous basis should be dumped. If oil is present on an intermittent basis, the condensate should be monitored with a turbidity meter or continuous organic analyzer. Aluminum sulfate with caustic soda or soda ash has been used to coagulate oil. The resulting aluminum hydroxide is then filtered in a pressure filter containing anthracite. Although not as effective as filters, mechanical separators are frequently used to separate oil from steam. In using such equipment, it should be kept in mind that oil cannot effectively be separated from *superheated steam*. The efficiency of the operation increases as the moisture content increases. Moisture can be increased in the steam by using unlagged pipe before the separators or by injecting condensate into the steam.

Iron and Copper.   In condensate systems, iron and copper corrosion are the result of carbon dioxide and/or oxygen. Ammonia must also be considered in the case of copper. The amounts of iron and copper present in the condensate should be low enough to meet the feedwater guidelines for boiler control. Normally, iron levels should be held below 50 ppb and copper levels below 20 ppb (Bumbard, 1982).

Control methods for iron and copper are the methods listed under carbon dioxide, oxygen, and ammonia, since these gases are the corrodants. Continuous flow of sample lines is the best method for iron and copper determinations. If this is not possible, the sample line should be run for about 24 h before a sample is secured. Iron and copper oxides are directly undesirable in condensate systems since they can plug lines and traps. In combination with filming amines, they form sticky deposits in both the condensate lines and the boiler. They also contribute to the formation of other deposits in boilers.

Inorganic Salts.   Salts in condensate can originate from boiler water carryover or leaks in the condensate system. Tests for boiler water constituents such as phosphate, alkalinity, or pH can indicate if the condensate salts originated from the boiler.

In the condensate system, carryover salts can interfere with the operation of filming amines. On the other hand, the high pH plus oxygen scavenger resulting from boiler carryover aids in preventing corrosion.

Salts originating from leaks of raw water into condensate are very undesirable. The water is saturated with oxygen, which aggravates condensate corrosion. A conductivity meter can be used to check for salt contamination. It can be used in conjunction with an automatic dump or alarm.

Cooling

To one who is not familiar with boiler and cooling water systems, boiler water systems would appear to present the greater number of problems; high temperatures are known to increase both scale and corrosion.

The actual conditions are the opposite of what might be expected. Cooling water systems present the larger number of problems, and these problems are the more difficult to correct. In consideration of the higher temperatures in boilers, the environment is changed. In virtually all boilers, the hardness and oxygen are removed from the feedwater. In high-pressure boilers, virtually all solids are removed from boiler feedwater. Economically, this cannot be done in cooling water systems. In addition, the designs and materials of construction are standardized to a much greater degree in boilers vs. cooling water systems. Finally, cooling water systems must contend with bacteria and algae that do not exist in boilers.

Judging from the number of variables involved in cooling

water scale and corrosion problems, it might appear very difficult to analyze problems. Yet, at times, the cause and solution to a problem is readily apparent. For example, stress corrosion cracking of brass takes place only in the presence of ammonia and stress; all other causes are eliminated from consideration. On the other hand, some corrosion problems include virtually all the variables that could possibly be involved.

9 Classifications of Cooling Water Systems

Corrosion Related to Type of System

Cooling water systems can be divided into three classifications: (1) once-thru, (2) closed, and (3) open recirculating. Certain characteristics of each system can assist in making at least a preliminary evaluation of a cooling water corrosion problem. In many cases, the treatment method is dictated by the type of system.

Once-Thru

Sulfate-Reducing Bacteria. If deep pits are found in a once-thru cooling water system, *sulfate-reducing bacteria* (SRB) should be considered as a possible cause. This type of bacterium is not restricted to once-thru systems, but conditions in such systems are often more favorable for its growth. Sulfate-reducing bacteria thrive in the absence of oxygen, in the presence of high soluble iron (>0.3 ppm), plus sulfate. Only about 20 ppm or more of sulfate are required. These conditions exist in many ground waters.

SRB can attack most metals, including stainless steel. Water velocities high enough to erode the establishment of slime deposits can assist in reducing the growth of SRB and other organisms. Stagnant areas, including crevices, should be avoided.

Since SRB can exist in all three types of systems (open, closed, once-thru), the subject is covered in detail on page 161.

In once-thru systems, intermittent chlorination is recommended to control sulfate-reducing bacteria. A free chlorine residual of 1 ppm should be maintained for two 1-h periods per day. If soluble iron is present in the raw water, polyphosphate or phosphonate will have to be fed to the water just before the chlorine is added, to prevent precipitation of iron.

Iron Bacteria. Although *iron bacteria* (Figs. 9-1–9-3) can be found in all three types of cooling water systems, they are most often found in once-thru well-water piping. A major type of iron bacterium, gallionella, thrives in high-soluble-iron waters (>0.3 ppm) and requires only low amounts of oxygen. Gallionella can attack metals by two methods (Tatnall, 1981). Directly, they concentrate chloride and iron ions. The resulting ferric chloride directly attacks the metal, producing pitting or general corrosion. The bacteria can also corrode metal through the formation of voluminous, reddish-brown, slimy deposits. This is turn allows for the development of differential aeration cells. Pitting can develop under the deposits.

Iron bacteria can be identified in a laboratory under a microscope. In the field, heavy deposits on pipe carrying high-soluble-iron water are a good indication that iron bacteria are present. Other oxidizing agents, such as oxygen and chlorine, can also form hydrated iron deposits. Treatment for iron bacteria in once-thru systems is the same as for sulfate-reducing bacteria; that is, chlorine fed twice a day for 1 h each time. Free chlorine should be held at 1 ppm plus. As with sulfate-reducing bacteria, polyphosphate or phosphonate must be fed

Figure 9-1.   Longitudinally split steel pipe showing by-products of corrosion induced by gallionella; corrosion products removed (courtesy of Buckman Laboratories, Inc., Memphis, Tenn.).

ahead of the chlorine to prevent precipitation of the iron. Above an iron content of 2 ppm, phosphonate should be used in preference to polyphosphate.

General Corrosion and Pitting.   Once-thru systems that use surface water often experience oxygen corrosion; both *general corrosion* and *pitting corrosion* can result. There is nothing unique in once-thru systems in this respect. However, in view of the large volume of water involved in once-thru systems, the treating methods are different from closed or open circulating systems.

Once-thru systems are often required to be of potable-grade. This further restricts the types of inhibitor that can be employed. Polyphosphates are commonly used inhibitors. The usual quantity fed is 4–8 ppm (PO_4). Zinc salts, 1–3 ppm (Zn), are often added with the polyphosphates to improve corrosion-

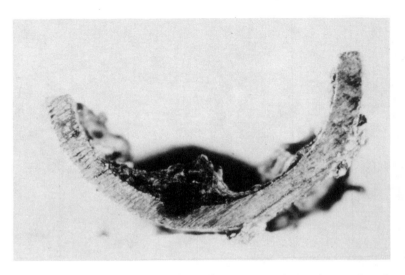

Figure 9-2. Cross section of steel pipe cut through a tubercle—gallionella-induced corrosion (courtesy of Buckman Laboratories, Inc., Memphis, Tenn.).

Figure 9-3. Inlet of heat exchanger heavily fouled with debris consisting mostly of corrosion by-products induced by growth of *Spherotilous* sp. and *Gallionella* sp. (courtesy of Buckman Laboratories, Inc., Memphis, Tenn.).

146

inhibiting properties. When soft-water conditions exist, silicates 5–10 ppm (SiO_2) are also added to supplement the polyphosphate.

Inhibition performance with these products is rated only as fair. In most cases, in once-thru systems, proper materials of construction must be used to resist corrosion. Although the choice and amounts of corrosion inhibitors are limited in once-thru systems, there is one advantage: the water quality is usually better than in open recirculating systems. There is no change in phase (water to vapor) as in an open recirculating water. Accordingly, corrosive salts such as sodium chloride and sulfate do not concentrate. Stagnant or low flow conditions should be avoided through design or operating conditions.

Closed

In a closed cooling water system, there is little loss of water, making it the easiest of the three bsic water systems to treat. As there is little loss, as much inhibitor can be added as necessary to control corrosion. Costs are only a minor factor. There are, however, a few problems that are peculiar to closed systems. Some of these are discussed below.

Nitrifying and Denitrifying Bacteria.   One common inhibitor that is used only in closed units is sodium nitrite. Bacteria can covert nitrite to nitrate or reduce it to ammonia or nitrogen. *Nitrifying bacteria* plate counts are not often determined in the field. But a rapid disappearance of nitrite, along with an increase in nitrate concentration, is strong evidence that nitrifying bacteria are present.

A rapid disappearance of nitrite and an increase in ammonia indicate the denitrification of nitrite. In either case, the nitrite inhibitor is lost, and the corrosion rate will probably increase. Nitrifying and *denitrifying bacteria* are difficult to control. When either type of bacterium is present, the best solution is usually to replace the nitrite with chromate or one of the newer types of inhibitors. Phosphonates, combined with silicates and tolytriazole, have been used with success in closed systems. Chromates, where permitted, are very effective in closed systems. Molybdates, in combination with tolytriazole and a *buffer* such as borax, are also being used successfully. If it is decided to remain with a nitrite inhibitor, a tin-quaternary ammonium biocide has often proved successful in controlling nitrifying and denitrifying bacteria.

Chromate-Ethylene Glycol. High chromates (250–1000 ppm) are often used as corrosion inhibitors in closed systems. They should not be used with ethylene glycol antifreeze solutions. The chromate will be reduced, changing color from orange to green. Since chromate is an oxidizing agent, it should not be used with any reducing agent. Nitrites and silicates have been used successfully in glycol systems.

Mechanical Seals—Chromates—Nitrites. Both chromates and nitrites have been blamed for the failure of mechanical seals in pumps (NACE, 1981). A report released by the National Association of Combustion Engineers (NACE) in 1981 concluded that the chromates did not attack mechanical seals when chromate levels were maintained below 500 ppm $CrO_4^=$. Nitrites below 4000 ppm NO_2^- caused no increase in seal wear. Satisfactory corrosion rates should be attained at these dosages.

Types of Inhibitors. Since corrosion inhibitors placed in closed systems remain there for a long period of time, care must be used in their selection. Inhibitors that have limited stability should not be used. This rules out the use of polyphosphates or phosphate esters, both of which hydrolyze.

Since closed systems normally have no major scale problems, they usually operate at higher pH values for greater corrosion protection. In view of the high pH, inhibitors such as zinc cannot be used since they precipitate. Nitrites and chromates are both effective and should be used in most systems.

Silicates or other proprietary treatments can be considered if chromates or nitrites are not permitted because of their toxic properties.

Open Recirculating

Of the three types of cooling water systems, only the open recirculating system represents a phase change. In passing over a cooling tower, a portion of the water vaporizes into the air. The heat used to vaporize the water is taken from the remaining water; this represents the major cooling. This system is also the only one that involves washing of air with circulating water. Since a number of contaminants are present in air, they account for several conditions that are unique to the open recirculating unit.

Drop in pH.   (1) *Ammonia:* Ammonia can enter a recirculating cooling water system through leaks in an ammonia plant, ammonia refrigeration plant, or other industrial processes (Schroeder and Landborg, 1976). Nitrifying bacteria, through the following reactions, convert ammonia to nitric acid.

$$2\ NH_3 + 3\ O_2 \rightarrow 2\ HNO_2 + 2\ H_2O$$
$$2\ NO_2^- + O_2 \rightarrow 2\ NO^-_3$$

Nitrifying bacteria plate counts are not usually determined in the field. A pH drop plus an increase in nitrate concentration in the cooling water are strong indicators that nitrifying bacteria are present. A test for nitrite is not recommended since the nitrite rapidly converts to nitrate; only small amounts of nitrite will be detected.

When ammonia is present in cooling water, the pH of the system should be determined after, not before, the water has passed through heat-exchanger equipment and is returning to the tower. The reason for this recommendation is that the pH can drop a full unit or more in passing through equipment. Severe corrosion can result from low pH. Every step, of course, should be taken to stop the ammonia leaks. Normal acid feed should be reduced or eliminated to compensate for the acid produced by the nitrifying bacteria. In some cases, it might be necessary to add soda ash for pH control.

(2) *Sulfur dioxide:* In many large cities, the air contains so much sulfur dioxide that acid addition to recirculating water is not required. Instead, the addition of soda ash is required to elevate the pH to a range of 6.5–8.0.

(3) *Carbon dioxide:* High concentrations of carbon dixoide are occasionally found in the air around commercial bakeries or other industrial plants. Low pH will result in the cooling water. Soda ash should be added to raise the pH. Caustic soda is not recommended for several reasons: it is difficult to handle, can cause sharp changes in pH, and can also precipitate many metal ions as the hydroxide.

Airborne Contaminants. Dust in the air can indirectly cause corrosion in a cooling water system. Washing of the dirt from air will result in mud deposits at various low-velocity parts of the system. Differential aeration cells can then develop, with pits forming under the deposits. The dirt can be either filtered through a side steam filter or held in suspension through the use of an organic dispersant such as a low-molecular-weight polyacrylate.

Types of Corrosion: Identification, Cause, and Prevention

The preceding section dealt with corrosion examples limited primarily to one of the three basic cooling water systems: (1) once-thru, (2) closed, and (3) open recirculating. The corrosion problems that follow can be found in all three types of systems although some may be found more often in one than the others. An analysis of the form of corrosion provides an excellent tool for identifying the cause and solution of many corrosion problems.

General Corrosion

This type of attack involves an even rate of corrosion over a wide range; general thinning of the metal is the end result. Usually, it is not of major importance in cooling water systems since the corrosion rates of materials of construction can be fairly well predicted under neutral operating conditions. Occasionally, however, overfeed of acid for pH control can result in general corrosion.

Attack resulting from oxygen under neutral conditions in heat exchangers is usually a mixture of general and pitting corrosion. Pitting corrosion is the more destructive. Often, insoluble corrosion products form differential aeration cells, and pitting follows. To prevent general corrosion, overfeeding of acid should be avoided. In addition, the proper metal for the given environment should be used. This is a general recommendation, but specific metals and environments are covered under separate headings. Proper inhibitors should be employed; these are also covered under a separate heading.

Pitting

In cooling water systems, pitting (Fig. 9-4) is the most destructive and difficult corrosion to control (Schroeder, 1972) since virtually any change in an energy level on a metallic surface can result in the start of pitting. Probably the best-known example is differential aeration, resulting from differences in oxygen concentration. Where deposits of any type are present, there will be little or no oxygen under the deposit, while the bulk of the solution will contain a much higher amount of oxygen. Corrosion in the form of pitting will take place under the deposit. Differential aeration cells are the most common type of pitting using mild steel, stainless steels, and aluminum. Halides, such as chlorides, aggravate the problem.

Pitting will often begin from slight imperfections on the surface of a metal. Usually, pits grow in the direction of gravity since the solution in the pit normally has a lower pH and

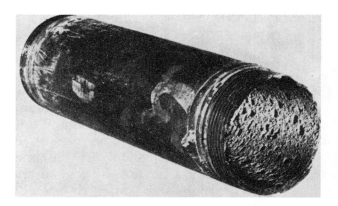

Figure 9-4.   Pitting corrosion.

higher salt concentration than the bulk solution. Accordingly, pits are more often found on the bottom side of horizontal piping. Other factors also account for pits being located at the six o'clock position of horizontal pipe. If debris is present, it will settle on the bottom side under low-velocity conditions. This is conducive to the development of differential aeration pits under the deposits. In addition, if piping is drained, it often tends to remain wet on the bottom side, encouraging corrosion in this area. Materials that are subject to attack by pitting will also be subject to *crevice corrosion* (see page 166), but the reverse is not always true.

Many energy levels can result in pitting. Included are differences in oxygen levels, metals *(galvanic corrosion)*, salt concentration, pH, velocity, metal surfaces, and temperatures. Metals that form strong oxide films to protect against corrosion are especially susceptible to pitting corrosion. Stainless steel and aluminum are two familiar examples. Localized high-chloride concentrations usually initiate such corrosion. Metals that depend on oxide film formation perform best when oxygen is present.

To avoid pitting corrosion, two general rules can be stated. (1) Keep all surfaces as clean as possible. Allow no deposits to form or, at least, hold to a minimum. The second rule is merely an extension of the first. (2) Avoid all differences in design and operation: stagnant areas, crevices, hot and cold adjoining areas, different velocities, etc. should be eliminated where possible.

Avoid metals that pit when pitting conditions prevail. Finally, employ inhibitors to hold corrosion to a minimum.

Oxygen pitting frequently takes place during shutdown periods in heat exchangers. During these periods, the water on the shell side should be full or completely drained. If full, high concentrations of oxidizing inhibitor, such as chromate (1000–2000 ppm) CrO_4, should be used.

Galvanic

This type of corrosion (Fig. 9-5) takes place when one of the general rules of corrosion prevention is broken, namely, that a system should be as homogeneous as possible. When two different metals are in contact in the same environment, the more active metal will corrode. Usually, the attack on the active metal will be close to the point of contact with the more noble metal. But there are exceptions. If the solution is highly conductive, the attack on the active metal may occur away from the junction of the two metals. Galvanic corrosion is not restricted to the direct joining of two metals. When the ions of a more noble metal are present in solution, they tend to deposit as the metal on more active metallic surfaces. As an example, if water flows through copper pipe, some copper ions will be in solution. If this solution then contacts zinc, it will deposit as metallic copper, and the zinc will go into solution. On the surface, a copper-zinc galvanic cell will develop, and additional amounts of zinc will be attacked in the form of pitting.

The initial classification of a metal as active or noble (galvanic series) cannot be depended on in the prevention of galvanic corrosion. Through the formation of films, the characteristics of metals can sometimes change. Zinc in contact with steel, for example, will be corroded up to about 140°F, protecting the steel surface. Above that temperature, zinc often becomes the cathode, and the steel is attacked.

Figure 9-5.   Galvanic corrosion in pump housing (copper-aluminum couple).

In actual practice, it is impossible to use the same metal for all applications. Nevertheless, steps can be taken to reduce galvanic corrosion to a minimum. When galvanic cells exist, the active metal should be used on the large area and the more noble metal on the smaller area. For example, brass valves should be combined with steel pipe, not the reverse. Or steel bolts can be used with aluminum plate, but not aluminum bolts with steel plate.

Listed here is the galvanic series:

Active, corroded end

Magnesium	Copper
Zinc	Bronzes
Aluminum	Copper-nickel alloys
Cadmium	Nickel
Mild steel	Inconel
Alloy steels	Stainless steel (passive)
Cast iron	Titanium
Stainless steels (active)	Silver
Lead	Gold
Tin	Platinum
Brasses	Noble end

Although this series is not foolproof, it should not be ignored. Metals should be selected that are similar in activity. Where possible, the metals should be electrically insulated from each other.

Galvanic corrosion is more severe in the presence of highly conductive solutions. Unfortunately, the plant operator usually has no control over cooling water quality. There are exceptions; the pH of recirculating cooling water can usually be controlled. Temperature can also affect galvanic corrosion rates but normally cannot be controlled.

Finally, noble metal piping should not be used ahead of more corrosive metals; do not use copper or copper alloy piping ahead of aluminum or zinc.

Corrosion inhibitors will not usually eliminate galvanic corrosion but will assist in reducing the severity of the attack.

Stress Corrosion Cracking

An especially destructive type of corrosion is stress corrosion cracking (Fig. 9-6). In this corrosion, a particular type of metal alloy exhibits a tendency to crack in a specific environment. Surface tensile stresses must also be present for cracking to occur. The stress can be either applied or residual. Cracking can often result in rapid destruction of equipment.

Figure 9-6.   Stainless steel stress corrosion cracking.

For *austenitic* stainless steels, the specific environment is chlorides plus oxygen. Caustic soda will also crack austenitic stainless steel but at higher temperatures than are found in cooling water systems. Caustic failures also tend to occur at a change of phase—from water to steam.

Copper alloys such as brass crack when under stress in the presence of ammonia or amines. Only very small amounts of ammonia have to be present.

Although not a major material of construction, aluminum alloys will also crack in the presence of chlorides and oxygen plus stress. Where stress corrosion cracking occurs, the type of attack should be confirmed by metallurgical examination.

When stainless steel fails from stress corrosion cracking, the cracks are usually many and branched. This type of crack is usually transgranular (without regard for grain boundaries). For brass or other types of copper alloys, the type of attack is ordinarily *intergranular* (along the grain boundaries).

Most often, general corrosion does not occur before stress corrosion cracking takes place; as a result, no advance warning of crack failure is given. If stress cracking conditions exist, measures should be taken to prevent cracks from forming. Preventive steps include: (1) eliminate the element in the environment that is causing the cracking; (2) when cracks have occurred, consider changing the type of metal in use; (3) reduce stress; (4) add or change inhibitor; and (5) reduce temperature.

In the case of austenitic stainless steels, it is often impossible to eliminate chlorides completely. Nevertheless, they should not be allowed to concentrate. Alternate wetting and drying have been shown to be more destructive than flooding. Lower temperatures assist in reducing the danger of stainless steel stress corrosion cracking, which generally does not occur below 150°F.

Oxygen is considered necessary for stress corrosion cracking of stainless steel. In some few cases, oxygen can be removed by the addition of sodium sulfite.

Stresses can be present from any number of sources: applied, thermal, welding, etc. In some cases, the cause is apparent and can be removed. When cracks develop in austenitic stainless steels, a change to ferritic stainless should be considered. These steels are virtually immune to stress corrosion cracking.

Brasses are susceptible to stress corrosion cracking in the presence of ammonia or amines. Unfortunately, only small amounts are required, and attack can take place at ambient temperatures. Leaks of ammonia are not necessarily present since the ammonia may originate from breakdown of various organic compounds. Where cracks occur, a change in alloy should be considered. Cracking usually takes place in brasses containing more than 20% zinc. Resistance improves as zinc is reduced. Cupronickel alloys are very resistant to stress corrosion cracking.

Where cracking occurs in the use of aluminum, a change to another metal should be considered.

Corrosion inhibitors can aid in the prevention of stress corrosion cracking. In the case of anodic inhibitors, however, it is very important that a sufficient amount be added.

Impingement

Soft metals like copper and aluminum are attacked in a corrosive environment in the presence of high velocities (Fig. 8-1). The attack is characterized by directional flow. Deep horseshoe or teardrop-shaped grooves are often present. No corrosion products are present in the grooves.

A number of factors influence the rate of attack, but velocity appears to be the major influence. If particles are present on the surface, the groove tends to be horseshoe-shaped. Otherwise, the grooves are likely to have teardrop shapes. When using copper pipe or tubes, the velocity should not exceed 4–5 fps. Other factors, which have less influence, are oxygen, pH, temperature, and water quality. Attack in aerated water increases with chloride content and decreased pH. The presence of large bubbles in the water also has an influence on impingement. Aluminum is another soft metal that one would expect to be subject to impingement attack. There is, however, little data available since the use of aluminum tubing is minor when compared to copper and copper alloys.

Since impingement is dependent on velocity, false corrosion readings can sometimes be given by coupons or probes. The corrosion measurements are often taken from racks where the velocity is regulated to the correct range. In the actual system, velocities may be quite different. As a result, corrosion measurements may be quite low while severe impingement actually exists in some parts of the system. One method of checking the corrosion measurements is to test for the metal ion level in the water. If the copper ion level, for example, is high, it can be expected that the corrosion rate is high, at least in some part of the system, despite corrosion measurements.

There are several steps that can be taken to reduce or eliminate impingement. Some steps can be taken immediately, while others require major changes in equipment. Some inhibitors, especially chromate, aid in reducing impingement. A newer type of inhibitor, a high-molecular-weight polymer, tends to reduce impingement by decreasing friction. It has been used primarily in closed systems (Hwa et al., 1972).

If velocity can be reduced, the impingement corrosion rate can be decreased. Where major changes in equipment can be made, the velocity can be diminished by enlarging the diameter of the tubes. A change in material of construction is also frequently made to eliminate impingement. As zinc content is increased in brass, the impingement resistance increases. Stainless steel and cupronickel alloys have high resistance to impingement attack.

Since the inlets to heat-exchanger tubing are frequently subjected to impingement attack, plastic *ferrules* are often installed at that point.

Cavitation

Impingement and *cavitation* are similar in that, in both forms of attack, mechanical forces destroy the protective outer surfaces on metals (Fig. 9-7). Corrosion also plays a role in cavitation,

Figure 9-7.   Cavitation pits on diesel liner (courtesy of Gordon and Breach Science Publishers, *Why Metals Fail*, Barer and Peters, 1970).

but cases of attack on nonmetallic materials have also been reported.

The large majority of impingement damage occurs on soft metals like copper and aluminum. Cavitation, on the other hand, most often occurs on, but is not restricted to, mild steel, which is not a soft metal and is, of course, the most widely used metal.

While impingement customarily results in grooving, the metals attacked by cavitation most often present a pockmarked appearance. Cavitation is primarily the result of high velocities and wide, sudden changes in pressure. These two factors, in turn, cause the formation and collapse of water vapor bubbles. Damage to the metal surface results. To varying degrees, corrosion also contributes to the damage.

In cooling water systems, cavitation is most often found on the suction side of pump impellers, pipe elbows, and the lining of diesel engine jackets. In the case of diesel jackets, high vibrations are responsible for the formation and collapse of bubbles.

Immediate measures to prevent cavitation are limited. The quality of the circulating water has some influence but can seldom be changed.

Chemical corrosion inhibitors have been recommended. Chromates are the most effective of these although nitrites and molybdates have also been used. Unfortunately, the recommended dosages are so high (6000–8000 ppm) that they can be considered only for closed systems (Veerabhadra et al., 1972). Most suggested solutions for cavitation are mechanical. Rubber coatings have been used to absorb shock. Air has been injected for the same purpose. Specifying smooth finishes on pump impellers has aided in reducing sites for bubble nucleation (Kirby, 1979).

Many hard-surface metals like stainless steel are quite resistant to cavitation and have been used to replace metals subject to cavitation in pump parts. Also, where pumps are involved, enough head should be maintained on the suction side to avoid large pressure differences. In piping systems, sharp turns should be avoided to prevent excessive turbulence. Velocities should be held below 10 fps. Diesel systems should be designed to reduce vibrations to a minimum.

Microbiological

It has been estimated that up to 33% of all corrosion in cooling water systems is induced by microorganisms. Most bacteria grow best between 68–122°F. Since microbiological corrosion is so important, it is considered separately here rather than as a division of pitting, which is the usual end result. The major types of attack follow.

Sulfate-Reducing Bacteria.   This type of corrosion was mentioned under Once-Thru Systems (page 143) since it is frequently found in systems of that type. Because the bacteria are *anaerobic*, they can live only where oxygen is absent. This does not mean that the bulk of the water contains no oxygen. Sulfate-reducing bacteria can thrive under deposits, where no oxygen is present, even though *aerobic* conditions exist in the main body of water.

Sulfate-reducing bacteria should be a major suspect when black corrosion products are found in deep pits under deposits. If this occurs on steel, the deposit should be checked with a magnet. A nonmagnetic deposit is strong evidence that sulfate-reducing bacteria are present. Even if the deposit is magnetic, sulfate-reducing bacteria may still be present since some black magnetite may be in the vicinity, along with ferrous sulfide (the end result of sulfate-reducing bacteria). As further evidence, if hydrochloric acid is added to the black deposit, hydrogen sulfide will be released if sulfate-reducing bacteria are present. Hydrogen sulfide is characterized by its rotten egg odor. A more sensitive test can be made with Lugol's solution. When the reagent is poured on a sample of sulfide scale, it will effervesce violently. This test is not specific for sulfate-reducing bacteria, but sulfides are rarely found in cooling water systems unless sulfate-reducing bacteria are present.

In appearance, pits resulting from sulfate-reducing bacteria are usually round at the outer surface and conical. When the corrosion products are washed from the pits, bright metal is evident.

American Petroleum Institute (API) culture bottles are available which contain 9 ml of the medium. A sample is obtained in a sterile syringe and injected through a rubber septum on the vial. The sample should be secured from the wall of the surface suspected of harboring sulfate-reducing bacteria. The vial is placed in an incubator at 37°C (98.6°F) and inspected daily for 10 days. The vial will turn black if sulfate-reducing bacteria are present.

Most of the major metallic materials of construction can be attacked by the bacteria. This includes copper, which is often believed to be immune to bacterial attack. Sulfate-reducing bacteria are found more often today since chromate inhibitors have been largely replaced with nontoxic inhibitors. Chromate above 50 ppm CrO_4 is an effective inhibitor of sulfate-reducing bacteria. Chlorine is also effective in preventing or controlling the bacteria at a dosage of 1 ppm free chlorine. Two daily 1-h feeding periods will usually prove effective. A number of proprietary biocides are also effective in curbing the growth of sulfate-reducing bacteria. The main difficulty is contacting the bacteria when they are growing under deposits. A preventive program is much more effective than trying to control a fouled system. If equipment is covered with slime, it may be necessary to disassemble the equipment and clean it with water blasts or sandblasting (Tatnall, 1981).

Less successfully, slime has been treated with air bumping, backflushing, steam sterilization, and large amounts of chemical dispersants.

Iron Bacteria (Gallionella).   Gallionella is not the only type of iron bacterium (Figs. 9-1–9-3) encountered, but it is a common one in cooling water systems. It thrives in high-soluble-iron waters and grows in low- as well as high-oxygen waters.

Gallionella corrodes steel by two methods. First, it oxidizes soluble ferrous iron and produces large volumes of hydrated iron deposits; corrosion develops underneath the deposits from *differential aeration* cells.

Second and more directly, gallionella concentrates ferric and chloride ions. The resulting ferric chloride then corrodes the steel. Gallionella should be considered as a possible corrodant whenever large amounts of reddish-brown deposits are present in high-soluble-iron waters. Its presence should be confirmed by direct examination under a microscope at 450–1000 magnifications (Lutey, 1980).

Gallionella can be controlled with chlorination at 1 ppm free chlorine or with proprietary biocides. If high levels of soluble iron are present, it will be necessary to add an iron-control agent such as polyphosphate or phosphonate during the chlorination period.

Nitrifying Bacteria.   This type of bacterium is normally not considered a possible cause of corrosion. Nevertheless, in open recirculating systems, the bacteria can contribute to corrosion through the formation of nitric acid. Refer to Drop in pH, page 149.

Nitrifying bacteria are very difficult to control. A better approach is to make every effort to stop ammonia leaks, which act as a nutrient for nitrifying bacteria. In addition, a regular treatment of biocides should be followed.

Dezincification

A distinctive type of corrosion known as *dezincification* (Fig. 9-8) occurs in brasses containing less than 85% copper. Selective removal of zinc leaves a spongy copper deposit, which has no structural strength. Prevailing opinion is that both zinc and copper are dissolved and that the copper is redeposited. Where attack is general, the corrosion is known as layer-type de-zincification. The deep-pitted type of corrosion is known as

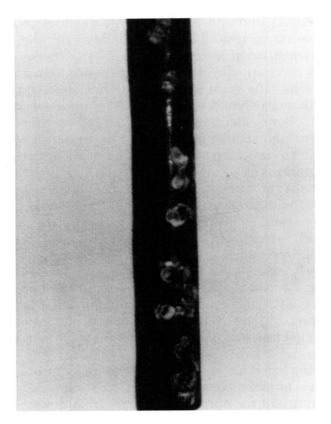

Figure 9-8.   Plug-type dezincification.

plug-type dezincification. Copper plugs frequently fill the pits, but many plugs are washed out of the pits. It is not unusual for the pits to penetrate the wall thickness completely. Both types of dezincification are readily recognizable by the reddish color of the redeposited copper in contrast to the yellow of the brass.

Within the range of zinc content in which dezincification occurs, the layer type seems to occur in the high-zinc-content brasses and in an acid environment. The plug type, on the other hand, is more prevalent in the lower-zinc-content brasses. It occurs most often in alkaline, neutral, or slightly acidic environments.

As a general rule, dezincification takes place in stagnant or low-velocity waters. Other contributing factors are increased temperatures, high dissolved solids, porous inorganic scale, and brackish waters (Uhlig, 1971, pp. 327–329).

On a short-term basis, the operating engineer or chemist can alleviate the problem only by making changes in the environment. In open recirculating systems, some changes can be made. These include changes in pH and total dissolved solids plus addition of scale dispersants. A more satisfactory and permanent solution is to replace the failed tubes with an alloy that will resist dezincification attack. Brasses containing less than 15% zinc are resistant to dezincification; resistance is also increased by the addition of an arsenic inhibitor.

Corrosion Fatigue

As the name suggests, corrosion fatigue is a combination of corrosion and alternating stresses. Unlike stress corrosion, a specific environment is not required. Virtually all natural waters are corrosive to some degree and play an important role in corrosion fatigue failures.

Corrosion fatigue has several distinguishing features that aid in its identification. Fatigue failures usually consist of only one major crack, whereas corrosion fatigue consists of many cracks in addition to the corrosion fatigue crack that fails. Corrosion fatigue cracks usually contain corrosion deposits, while fatigue cracks are usually clean although they do occasionally contain deposits.

The type of environment must also be considered (Schwartz, 1982). Conditions for alternating stresses must be present in addition to corrosive surroundings. The stresses can be vibrations, alternating thermal stresses, or water hammer. Often vibrations result in tubes rubbing against each other; evidence of wear are present (Evans, 1960). Under the microscope, corrosion fatigue cracks have blunt ends and few branches; stress corrosion cracks, in contrast, are branched and have fine ends. Also, corrosion fatigue failures are most often transgranular, while stress corrosion cracks may be either intergranular or transgranular.

Oxygen, temperature, pH, and ion concentration all have an effect on corrosion fatigue. Some of these factors can be controlled. Corrosion inhibitors such as chromates have been used to control corrosion fatigue. It is important that a sufficient amount be used or the attack will be more severe; 200 ppm as Na_2CrO_4 has been mentioned as the amount required to prevent corrosion fatigue. This is a very large amount to be maintained in an open recirculating system. It is also advisable to use a cathodic inhibitor such as zinc in combination with the chromate.

The most effective measures customarily require capital expenditures. One step is to use more resistant metals. Where corrosion fatigue is concerned, corrosion resistance is more important than structural strength (Uhlig, 1971, p. 147). Copper, stainless steel, and nickel are all more resistant to corrosion fatigue than carbon steel. Steps should be taken to reduce the alternating stress, be it vibration, thermal, water hammer, etc. Organic coatings have also been used successfully to resist corrosion fatigue. Coatings containing inhibitors are the most effective.

Crevice Corrosion

One of the basic rules in the prevention of corrosion is to maintain all conditions as homogeneous as possible. Crevice corrosion (Fig. 9-9) is a prime example of the damage that results when this rule is broken.

This type of corrosion can be recognized by the localized attack that results, the location of the attack, and the type of metal employed.

For crevice corrosion to occur, stagnant water must be in

Figure 9-9.   Galvanized steel crevice corrosion.

contact with a bulk solution through a small crevice; a concentration cell develops. It was once thought that the difference in oxygen concentration between the crevice and the bulk solution was solely responsible for the attack. It is now thought that the major part of the attack results from chloride concentration and low pH (Keys, 1976). Passive film metals like stainless steel are more susceptible to this type of attack than active metals such as steel or inactive metals such as copper.

To prevent crevice corrosion, every step possible should be taken to avoid the development of stagnant areas. All equipment should be designed to drain. In flanges, fibrous gaskets should be avoided; products such as Teflon are preferred. During long shutdown periods, packing should be removed from pump stuffing boxes. Threaded pipe fittings should be tight and caulked. Welded joints are preferred to threaded ones. Tubes welded to tube sheets are also preferred to rolled tubes.

An all-inclusive list cannot be compiled. But the operating engineer can use his own ingenuity in making certain that crevice conditions are avoided (Fontana and Greens, 1967). As an extra precaution, equipment should be inspected frequently.

There is often a delay of six months or a year before crevice corrosion is evident. Once started, corrosion is rapid. In some cases, the water quality can be changed; water containing low chloride is preferred. Dispersants can also assist in keeping surfaces clean. This can aid in preventing crevice corrosion under mud deposits. Where conditions permit, higher velocities also assist in maintaining clean surfaces. Corrosion inhibitors are of limited assistance in preventing crevice corrosion since the inhibitor cannot contact the area being attacked.

Intergranular Corrosion

When the 300 series stainless steels are welded, the temperature in the immediate area reaches 900–1500°F. Carbon at these temperatures diffuses to the grain boundaries, where it combines with chromium to form chromium carbide. This depletion of chromium makes the stainless steel less resistant to corrosion; galvanic cells are also set up. Intergranular cracking in the weld area is the result (Fig. 9-10).

The movement of carbon to the grain boundary is a time-temperature reaction. One method of avoiding the problem is to use arc welding instead of gas torch welding. With arc welding, the necessary temperature is reached in a shorter period of time.

Solution quenching and the addition of stabilizers are methods used to prevent intergranular corrosion. Unfortunately, these methods are not readily available to the operating engineer. The use of lower-chloride water and the addition of corrosion inhibitors will be of some benefit. Overall, intergranular corrosion represents a defect in the welding procedure, and the solution should start at that point.

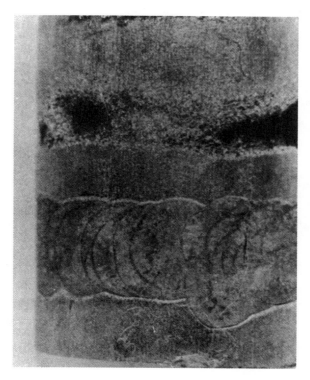

Figure 9-10.   Intergranular corrosion (Courtesy of NACE, Houston).

10 Corrosion-Related Factors

Corrosion Related to Physical Factors

Design

The very best inhibitor will not be effective if it fails to contact the metal; in addition, excellent-quality cooling water may still be very corrosive if it is allowed to concentrate. These are but two examples that illustrate the importance of good plant design. There are many variations in the sizes and shapes of plants, but some basic rules apply to all plants in the effort to control corrosion.

In daily operations, it may not always be possible to change plant design. Nevertheless, plant operations can sometimes be modified to fit the design, and minor plant design changes can sometimes be made in an effort to control corrosion.

Homogeneity of Equipment. If a cooling water system is designed with no planning given to differences in metals, velocity, oxygen concentrations, etc. it will almost assure corrosion problems. The obvious solution is to make all parts of the system as homogeneous as possible. Metals in contact should be similar in activity, sharp changes in velocity should be avoided, etc. Crevices of all types should be avoided. Where gaskets are employed in flanges, they should be inert and nonporous. The gaskets should not extend into the line of flow. Corrosion is also less likely to occur on a smooth than a rough surface. Briefly, all differences in a system should be held to a minimum.

Stagnant Areas.   When pools of water are allowed to remain in any part of cooling water equipment, the prospects for corrosion increase. In addition, the concentration of oxygen in the stagnant water varies, setting up chances of differential aeration cells developing. Also, many corrosion inhibitors are not effective in stagnant water.

A survey of the plant should be made to make certain that all piping, tanks, heat exchangers, etc., are designed to drain. If a section of pipe cannot be drained, it should be filled; the poorest condition is a partially filled line.

Velocity.   This factor is extremely important in corrosion-control planning. Systems vary; the best velocity for one metal may be incorrect for another. Copper, for example, will suffer impingement corrosion at high velocities. The top velocity for copper is about 4–5 fps. Mild steel usually operates at 5–7 fps; 316 stainless steel can withstand velocities up to 15 fps. High velocities are involved in cavitation and impingement problems. Water hammer can also develop at high velocities. To assist in preventing water hammer problems, the water velocity should be maintained below 10 fps.

If velocities are low in an open recirculating cooling water system, the chances that deposits will form are increased. Low velocities also mean that less corrosion inhibitor reaches the surface. In piping, high-velocity problems usually occur at sharp turns. Smooth, gradual bends should be made. If a specific water flow must be maintained for process reasons, two corrective steps can be taken: the construction material can be changed, or the diameter of the pipe can be changed. In shell and tube exchangers, both high- and low-velocity problems can be encountered. High velocities often cause damage to the inlet tubes on the tube side. This can be corrected by the insertion of plastic ferrules in the tube entrance. For the long term, it should be kept in mind that deep-water boxes assist in preventing velocity erosion problems.

On the shell side, low water velocities can cause deposition of suspended matter in the baffle area. Corrosion can then occur under the deposit. Modifications in baffle design can reduce the amount of deposit; sludge dispersants can also assist in preventing deposits.

Stresses. In the use of metals that suffer stress corrosion cracking, precautions should be taken to assure that no applied stresses develop. Stresses frequently occur in the installation of piping, pumps, etc. Another type of stress—resulting from vibration—is involved in corrosion fatigue. Where corrective steps can be taken, applied stresses and vibrations should be held to a minimum.

Galvanic Contacts. This type of corrosion (Fig. 9-5, Table 2) is covered in Chapter 9, page 153 (see also the galvanic series, page 155). It is also involved in the design of equipment against corrosion. Metals in contact should be similar in activity. Also, the more active metal should be used in the larger area. Brass valves, for example, with steel pipe are acceptable; steel valves with brass piping represent poor design.

Electric insulating gaskets are frequently used to prevent metal-to-metal contact. Two precautions should be taken. It must be made certain that the gasket is an insulating product. Problems also develop at times because all parts are not insulated. An example is a flange with a gasket to separate and seal the flange heads. If the bolts are not insulated also, the flange will not be electrically insulated.

Galvanic corrosion can also take place even though the main components are not in direct contact. In a zinc-copper system, for example, copper ions will deposit as the metal and zinc will go into solution. In a once-thru system, zinc can be used before copper but not the reverse.

Sample Lines—Inspections.   Sample lines should be installed in areas where corrosion can develop. The water quality in one area is not necessarily the same in a different part of the system. Also, provisions should be made for frequent inspections so that corrosion can be detected before any major damage has taken place. Where deposits occur, corrective action can be taken with no resultant damage; but where corrosion develops, the damage is permanent. For this reason, monitoring of corrosion must be thorough and frequent.

Costs.   Economics must be included in any system design. For example, in a closed system, little water is lost so that high concentrations of corrosion inhibitor can be used. As a result, expensive materials of construction are not required in many cases.

At the opposite end, high concentrations of corrosion inhibitor cannot be used in once-thru systems because of cost considerations. Instead, more resistant and usually more expensive materials of construction are required.

Corrosion Related to Inhibitor Employed

Cooling water inhibitors are intended to prevent corrosion. But, in many cases, they fail to achieve their intended result and may create other problems. Problems with inhibitors usually arise through incorrect applications, high and low dosages, incompatibility with other water treatment chemicals, and scale formation. In all cases, the inhibitor must be evaluated in light of individual plant conditions.

The following are the most widely used inhibitors.

Chromates. Alone, or combined with other inhibitors, the chromates are recognized as the most effective cooling water inhibitor. Unfortunately, chromates are toxic, and their use is prohibited in comfort cooling towers. Their primary use today is in closed systems, where water loss is minimal, and in industrial open recirculating systems, where their use is permitted. In some large plants, the corrosion conditions are severe enough to require continued use of chromates in spite of their toxic properties. Under such conditions, chromates can continue to be employed by removing the chromate before it is discharged from the plant. Several methods have been employed for this. The reduction method is by far the most widely used. The pH of the discharge water is lowered to 3.0–3.5 through the addition of sulfuric acid. A reducing agent is then added to convert the hexavalent chromate to trivalent chromium. Sulfur dioxide is the most widely used reducing agent although ferrous sulfate and sodium sulfite have also been employed.

Following the reduction, lime or other alkaline materials are used to raise the pH to about 8.5. Some plants have used the blowdown from cold-lime treaters as the source of alkalinity. Electrolytic reduction and ion exchange have also been used. The electrolytic method is effective, but the economics of the operation depends on power costs. The ion-exchange method has been employed in only a few plants. Maintenance problems have been reported as high, with restricted life of the ion-exchange resin.

Chromates are oxidizing anodic-type inhibitors. When they are used, it is important that the dosage be sufficient to cover all anodic areas. If this is not done, pitting can result. In an effort to reduce toxicity, "low-chromate" type of treatments have been introduced. This involves using small dosages of chromate in combination with other inhibitors. In aggressive waters, the reduced chromate levels have sometimes resulted in pitting corrosion problems.

The amount of chromate used varies considerably. In open

recirculating systems, the quantity can vary from 2 to 30 ppm CrO_4 when used in combination with other inhibitors like zinc or polyphosphates. Varying operating conditions, which are covered elsewhere in this section, determine the required dosage.

In closed chill water systems, the recommended dosages are not a large factor since there is little water loss. Dosages are usually held below 500 ppm chromate (CrO_4) to protect mechanical seals.

For refrigeration brine systems, the recommended dosages are 125 lb of sodium dichromate per 1000 ft^3 of calcium brine (1300 ppm CrO_4) or 200 lb of sodium dichromate per 1000 ft^3 of sodium brine (2246 ppm CrO_4).

Since chromates are strong oxidizing agents, they should not be used with reducing agents. Carbamate-type biocides are an example of one such reducing agent. At times, attempts are made to mix chromates with organic chemicals in proprietary products. The effort almost always fails. If a chromate-treated cooling water is green instead of the familiar orange-red, it can be concluded that the chromate has been reduced and is no longer effective.

Polyphosphates—Phosphates.   With the declining use of chromate treatments, the polyphosphates have come into wider use, especially in combination with other products. Polyphosphates cannot be separated from orthophosphates since, to some degree in all systems, the polyphosphates revert to orthophosphates. Reversion is one of the main limitations of polyphosphates. The end product—orthophosphate—tends to form scale with calcium. For this reason, dispersants are usually combined with polyphosphates; they tend to prevent scale formation. Polyphosphates have other limitations. Under the very best of conditions, they are not as effective as chromates. As a result, corrosion may develop with the use of polyphosphate treatments where no problem existed with chromates.

Other items must be considered when polyphosphates are used. Oxygen is essential in their use. Accordingly, polyphosphates must be used in moving waters; they are not effective in stagnant areas. For this reason, they are not very effective on the shell side of heat exchangers. Stagnant pools tend to accumulate around baffles. Calcium ion improves the inhibiting performance of polyphosphate but is not absolutely necessary for satisfactory results. Phosphates also have the limitation of encouraging the growth of algae and bacteria. This makes the control of slime more difficult than when chromates are used.

Polyphosphates are used in low dosages (3–8 ppm) in potable once-thru systems because they present no toxicity problems. Corrosion protection at that dosage is minimal. But the water being treated is not as corrosive as open recirculating systems since the water is not being concentrated.

In open recirculating systems, polyphosphates have been effectively combined with phosphonates. This inhibitor combination is widely used and quite effective. The usual reversion problems exist, but the phosphonates tend to prevent deposition. Because of the reversion problem, polyphosphates are not used in closed recirculating systems.

Nitrites.   For closed systems, nitrites are effective inhibitors. Their main disadvantage is their susceptibility to microbiological attack. Depending on conditions, the nitrites can be oxidized to nitrates by bacteria or reduced to ammonia or nitrogen. Bacteria plus oxygen convert nitrite to such an extent that it cannot be used in open recirculating systems. Required dosages are high enough to rule out nitrites for use in once-thru systems.

Nitrites should not be used at a pH lower than 6.5 or where the pH may become low locally. Nitrites do not require oxygen to perform as corrosion inhibitors but are more effective when it is present.

Under the very best of conditions, the nitrites are not as effective as chromates—the product the nitrites usually replace in closed systems.

If biocides do not succeed in controlling the decomposition

of nitrite, a change to another inhibitor should be considered. In a number of cases, reports have been favorable regarding the use of quaternary-tin biocides to control nitrogen-related bacteria.

Nitrites have no inhibiting effect on brass and should not be used in such systems without the presence of a copper inhibitor (Butler and Ison, 1966). If the nitrites decompose to ammonia, they can be destructive to copper alloys. Nitrites in closed systems should operate on the alkaline side, but the pH should be held below 9.0. This is especially true if aluminum is present.

Since chromates in closed systems are known to have an adverse effect on carbon seals, it should be noted that a 1981 NACE report indicated no adverse effect from sodium nitrite concentrations up to 4000 ppm as NO_2.

Dosages in closed systems vary from 500 to 1500 ppm as $NaNO_2$.

Silicates.   The chief attraction of silicates is that they are nontoxic; they are also quite stable and are not attacked by bacteria. Unfortunately, silicates are not effective inhibitors in most cases. They are used primarily in closed and once-thru systems. In virtually all cases, they are combined with other inhibitors.

Silicates should not be used in hard waters. They are not as effective as in soft waters, and they frequently cause scale problems in hard waters if temperatures are elevated.

Tolytriazole (TTA)—Benzotriazole (BZT)—Mercaptobenzothiazole (MBT).   All three of these chemicals are used as copper inhibitors.

All of these products are affected by chlorination. Ordinarily, tolytriazole and benzotriazole are effective in the dosage range of 1–2 ppm. When chlorine is present, the dosage should be increased to about 5 ppm. This applies under continuous chlorination conditions. The dosage rate for mercaptobenzothiazole is also in the range of 1–2 ppm in the absence of chlorine. It should not be used in the presence of chlorine unless the chlorination is intermittent (Nathan, 1973). Unless zinc is present, MBT also interferes with the inhibition of steel by polyphosphates. This interference is not exhibited by BZT or TTA.

Zinc.   This cationic inhibitor is very seldom used alone. It is usually combined with chromate, polyphosphate, or phosphonate. The major problem with the zinc ion is that it precipitates in high-pH areas. Its use should be restricted to a pH of 8.0. The dosage of zinc ion should also be held below 2 ppm to control precipitation.

Phosphonates prevent zinc ion from precipitating at high pH values (e.g., 8.5) if the water is soft. Conversely, if approximately 50 ppm of Ca are present, the phosphonate preferentially retains the calcium, and zinc precipitates. Under these conditions, it is necessary to use pH control.

It has been claimed that phosphincarboxylic acid will prevent the precipitation of zinc in the presence of high-pH and high-hardness waters (Marshall, 1981).

Molybdates.   The chief attraction of molybdates is that they are nontoxic. Like silicates, they are not as effective as other inhibitors in wide use such as chromates or polyphosphates.

Initially, high dosage requirements, when used alone, restricted the use of molybdates. The large majority of products now use molybdate at low dosages, combined with other inhibitors such as zinc.

Oxygen is required for molybdate to function. This restricts its use where circulation is poor. The necessary oxidation function is sometimes provided by combining nitrite with molybdate.

Phosphonates are known primarily as scale inhibitors. They have mild corrosion-inhibiting properties when used alone; combined with zinc or polyphosphates, the corrosion-inhibiting properties are excellent. High phosphonate (>20 ppm) dosages should not be employed because the phosphonates attack copper. For this reason, it is advisable to employ copper inhibitors when phosphonates are used, regardless of the phosphonate dosage.

When continuous chlorination is used, aminomethylene phosphonate (AMP) should not be used since it decomposes. Hydroxyethylidene diphosphonate (HEDP) is relatively stable in the presence of chlorine.

Corrosion Related to Material of Construction

Cooling water corrosion problems cover many variables, including design of equipment, inhibitors, and quality of water. None is more important than the selection of metal in plant construction. If an improper selection has been made, two courses of action are available to control corrosion in the system. The metal in use can be replaced with the proper material which, for various reasons, may not be possible. The other choice is to change the other corrosion factors to suit the metal being used. It may be possible to change the inhibitor, lower the temperature or velocity of the water, or use a better-quality water.

Characteristics of the most widely used metals follow.

Mild Steel

This metal is the most widely used of all metals. Its resistance to corrosion may be rated from fair to poor. The wide use of steel results from its low cost compared to other metals. Also, its structural properties and ease of fabrication add to its attractiveness. The optimum pH of water solutions for steel is 11.5. Mild steel is subject to corrosion of the following types.

General. This type of corrosion usually takes place in low-pH waters. Oxygen corrosion of steel often starts as general corrosion but usually changes to pitting after deposits form.

Pitting. The resistance of mild steel to pitting is relatively poor. Pits most often develop under deposits.

Impingement. Mild steel can be attacked by this type of corrosion but is more resistant than copper.

Cavitation. This type of corrosion (Fig. 9-7) can affect mild steel. Stainless steel is more resistant.

Galvanic. Mild steel can form galvanic cells with other metals and is most commonly coupled with copper and copper alloys. When it is used with copper metals, the larger area should be mild steel.

Corrosion Fatigue. Mild steel can be attacked by this form of corrosion. Copper, stainless steel, and nickel are more resistant to it.

Microbiological Corrosion. Mild steel is readily attacked by sulfate-reducing bacteria and other forms of bacteria. Most other metals are also subject to attack.

In addition to types of corrosion, high temperatures and salt concentrations increase steel corrosion rates. Soft water also increases corrosion rates in comparison to hard water.

Stainless Steels

Stainless steel has excellent corrosion resistance, but it does have a few limitations. Only the austenitic stainless steels (304, 316, and 317) are covered here because they are more widely used. New ferritic stainless steels have come into specialized use because they are resistant to chloride attack.

General. Stainless steels have excellent resistance against this type of corrosion.

Pitting.   All surfaces of stainless steel must be kept clean. If this is not done, chloride pitting can result. Increased temperatures aggravate attack.

Impingement.   Stainless steels exhibit good resistance.

Cavitation.   Resistance is higher than mild steel.

Galvanic.   Stainless steels can enter into galvanic cells. It is usually the more noble metal.

Stress Corrosion Cracking.   This corrosion (Fig. 9-6) is probably the greatest weakness of stainless steels. Stresses should be avoided and surfaces kept clean. In order to avoid chloride buildup, stagnant areas should not be allowed to develop. This type of corrosion in stainless steels is not common below 150°F.

Microbiological Corrosion.   Much as with other metals, stainless steels can be attacked by bacteria.

Crevice Corrosion.   All crevices should be avoided when stainless steels are used. High chlorides aggravate the problem.

Intergranular Corrosion.   This type of attack (Fig. 9-10) occurs as a result of poor welding practices. Refer to Intergranular Corrosion, page 168.

Copper-Copper Alloys

The commonly used forms of copper-bearing metals in water treatment service are copper, brass, and copper-nickel. Selection of the proper material depends on relative cost, corrosion resistance, and mechanical strength. Recognition of these properties aids in both selection and problem solving involving use of the metals. The best-known form of brass is 70-30 (Cu-Zn). Copper and brass, between them, constitute about 90% of the metals containing copper. Copper-nickel alloys (70-30 and 90-10) make up most of the remaining 10%.

Copper and its alloys have overall good resistance to aqueous corrosion. There are certain guidelines in its use that should be followed.

General. Copper tubing shows good resistance to general corrosion in middle pH ranges. In the absence of oxygen, it shows good resistance to acids but, unfortunately, oxygen is usually present. Oxidizing acids such as nitric acid also attack copper. High chlorides increase attack.

The other type of general corrosion is by ammonia in the presence of oxygen and in the absence of stress.

Brass exhibits about the same degree of attack as copper under the same conditions. Copper-nickel alloys are much more resistant to attack by acids or ammonia. While 90-10 copper-nickel alloys exhibit reduced ammonia attack, 70-30 copper-nickel alloy is virtually immune to attack.

Galvanic. This type of corrosion often involves copper. It is customarily the more noble metal. Brass is less noble than copper; copper-nickel alloys are more noble than copper. Contact with zinc and aluminum should be avoided.

Pitting. Copper and brass pit under various conditions. Copper-nickel alloys are more resistant. Chlorine in soft water causes pitting. High carbon dioxide levels in the presence of oxygen in soft, medium- to high-dissolved-solids warm water can also pit copper and its alloys. Water containing manganese also pits copper.

Hard waters most often do not corrode copper or its alloys.

Since copper is a relatively stable metal, it is not affected to a great degree by differences in oxygen concentration. Nevertheless, copper tubing often pits in stagnant waters under deposits. Off-line systems should be flushed to remove such deposits.

Stress Corrosion Cracking.   This type of corrosion is one of the major problems in the use of brass. In the presence of ammonia, stress, and oxygen, cracking can develop. Copper is immune to this type of attack, while copper-nickel alloys are highly resistant.

Ammonia should be avoided in the use of copper alloys. Stresses should be eliminated when possible. Also refer to Stress Corrosion Cracking, page 155.

Impingement.   The use of unalloyed copper is severely limited by impingement (Fig. 8-1). Copper is a soft metal and suffers serious corrosion damage at high velocities or turbulence. Alloying with zinc in brass increases resistance to impingement. Copper-nickel alloys are very resistant to impingement.

Cavitation.   Copper and its alloys can be attacked by cavitation, as can most metals.

Microbiological Corrosion.   Copper is supposed to be somewhat resistant to bacterial attack, but corrosion from this source is not uncommon. Resistance appears to be greater in alloys with high concentrations of copper. Nevertheless, sulfate-reducing bacteria and other types of bacteria have been found on copper and its alloys.

Dezincification.   This type of corrosion is found in brasses. Resistance increases in brass with decreasing amounts of zinc.

Corrosion Fatigue.   This type of attack is found among copper and copper alloys. Resistance varies considerably, depending on the alloy in use.

Crevice Corrosion.   Although copper and its alloys can be attacked by crevice corrosion, they are more resistant than most metals.

Galvanized Steel

The corrosion resistance of galvanized steel is relatively good for ambient-temperature waters in the neutral-pH range. High temperatures should be avoided. It should probably not be used in recirculating cooling water systems employing acid addition for pH control; at some stage, too much acid will be added, and the zinc coating will be removed.

Also, all contact between galvanized steel and copper metals should be avoided. Performance is satisfactory in once-thru systems if the water first contacts galvanized piping before contacting copper lines. The reverse situation results in severe pitting of galvanized steel.

Where mechanical factors are involved in the corrosion process, the resistance of galvanized steel is similar to that of mild steel.

Structural members of air conditioning cooling towers are often made of galvanized steel. These sections often corrode through the concentrations of salts in wet-dry areas. Once damage has occurred, it cannot be corrected by water treatment. If possible, the affected sections should be sandblasted to white base metal. Then zinc-rich primer should be applied, followed by two coats of epoxy enamel. As an alternative, apply three coats of epoxy enamel.

Titanium. Along with its alloys, titanium exhibits excellent corrosion resistance. Unfortunately, it is expensive, and its use is restricted for that reason. Titanium is not subject to galvanic attack but can adversely affect other metals (Moore, 1979).

Aluminum. Because its corrosion resistance is relatively poor, aluminum piping, etc. is not widely used in industrial plants. It is most often used in oxidizing environments and neutral-pH solutions.

Corrosion of aluminum is aggravated by high temperatures and chlorides. In addition, aluminum is prone to deposit corrosion, a form of differential-aeration corrosion. Also, all contact with copper and its alloys should be avoided. On the other hand, metal-to-metal contact is relatively good for aluminum-stainless steel and aluminum-steel couples.

Corrosion Related to Water Quality

In addition to all other factors that have to be considered in cooling water corrosion, water quality must be included. It is not uncommon for a corrosion inhibitor to perform quite well in one water and fail in another. In some instances, it is possible that a choice can be made between one water source and another. More often than not, however, there is only one water supply available, and operating conditions must be able to handle the quality of that particular source. The properties of the various salts and operating conditions follow. The term "water quality" has been used in the broader sense to include temperature and dissolved gases.

Chlorides. In most cooling water corrosion cells, chlorides are involved. Other factors being equal, it can be assumed that the higher the chloride content, the more corrosive the water. At the present time, reverse osmosis is the only pretreatment that reduces chlorides. Although the use of reverse osmosis is increasing, it is not widely used at present for cooling system makeup.

When pits or cracks occur on stainless steel or other metals, chlorides are always suspect. A quick field test is to wash the pitted area with demineralized or distilled water. Add silver nitrate solution to the wash water. Cloudiness indicates the presence of chloride.

If chloride levels are high enough to cause severe corrosion, they can be controlled in open recirculating systems by limiting the *cycles of concentration.* Corrosion from chlorides can also be controlled by increasing the amount of corrosion inhibitor or changing to a more effective inhibitor.

Sulfates. Sulfate alone is not supposed to influence the corrosion of aluminum. In combination with chlorides, it increases the corrosion of aluminum (Godard et al., 1967). High sulfates increase the corrosion rates of virtually all other metals. Not only is sulfate present in most raw waters but, in many recirculating cooling waters, the sulfate content is increased through the addition of sulfuric acid. The acid is used for pH control.

Increasing blowdown from recirculating cooling water is about the only practical method of reducing sulfate content. Where sulfates are contributing to high corrosion rates, increased inhibitor dosages or a change of inhibitor can also be considered.

Oxygen. Oxygen serves a dual purpose. It is involved in many types of corrosion. Conversely, many metals depend on oxide films for their corrosion resistance. In addition, many corrosion inhibitors such as polyphosphates depend on oxygen for their effectiveness. In daily operation in cooling water systems, there is virtually no control over the oxygen content. Yet knowing the oxygen content makes the selection of the proper inhibitor and material of construction easier.

Bacteria. Metals are attacked by bacteria either directly or indirectly through the formation of slime. The best solution is prevention. Chlorine or other biocides should be used to maintain control. Refer to pages 201–204.

Gases. The most common gases involved in cooling water corrosion are ammonia, sulfur dioxide, carbon dioxide, and chlorine. In view of its importance, oxygen has been covered separately above.

Ammonia is best known for its attack on brass, resulting in stress corrosion cracking. Nitrifying bacteria can convert ammonia to nitrites and nitrates. This reaction lowers the pH, which can also corrode metals. The only defense against corrosion caused by ammonia is to eliminate the ammonia contamination. In the case of stress cracking, the replacement of brass can be considered. The actual replacement metal will depend on the specific application.

Sulfur dioxide is usually found in polluted air regions. The end result is a lower-pH water. Soda ash is used to correct this condition.

Occasionally, high carbon dioxide levels are found in industrial areas. As with sulfur dioxide, low pH levels result. Soda ash can be used to raise the pH.

Chlorine is present in many cooling water makeup supplies. The usual levels (0.5–1.0 ppm free chlorine) do not generally corrode most metals. Higher levels can result in pitting, especially in soft water. Some organic cooling water corrosion inhibitors are oxidized by low levels of free chlorine.

Temperature.   Increased temperatures generally result in higher corrosion rates of all types. An exception is aluminum. Up to 100°F, the rate of pitting on aluminum decreases (Godard et al., 1967).

Most often, temperatures cannot be controlled, but the need for an effective inhibitor, sufficient velocity, etc., must be recognized to compensate for the increased temperature.

Suspended Solids.   The presence of suspended matter in cooling water can increase impingement-type corrosion. *Suspended solids* may also deposit in low-velocity areas and create differential-aeration cells. Pitting can result.

In-line filters or various types of pretreatment can be used to lower the suspended-solids level. Various polymers assist in holding solids in suspension.

pH.   pH ranges of 7.0–9.0 have little influence on the corrosion rate of cooling waters. If, for some reason, such as pollution, the pH is lowered into the acid range, increased corrosion can be expected. The solution lies in determining the cause of the low pH and correcting that condition.

Iron and Manganese. Waters containing high iron levels (>0.3 ppm) promote the growth of sulfate-reducing bacteria and iron bacteria (gallionella). Both can cause pitting corrosion. In addition, hydrated iron deposits can form. Pitting can then develop under the deposits.

Manganese oxide can deposit on copper in soft water, resulting in pitting corrosion.

Iron and manganese can be either removed before they enter a system or controlled through the addition of polyphosphate or phosphonate. Chlorine, added after the dispersant addition, can control the bacteria. Customarily, 1 ppm of free chlorine added for 1 h twice a day will suffice.

Bicarbonates. In combination with other salts, bicarbonates usually increase corrosion. Normally, in large recirculating systems, the bicarbonates are controlled through the use of sulfuric acid.

Calcium. Calcium waters are more aggressive to aluminum than soft water. The reverse is true of the majority of other metals (Godard et al., 1967). For the most part, raw waters contain a sufficient amount of calcium to aid in controlling corrosion. If they do not, the importance of including a cathodic inhibitor such as zinc is increased.

Corrosion During Shutdown Periods

A large amount of boiler corrosion takes place during shutdown periods. In a like manner, if no preparations are made, severe corrosion problems can take place in cooling water systems during shutdown. Ordinarily, corrosion takes the form of pitting. The problem is less complicated with boiler systems since equipment layout is more standardized. Cooling systems can include a variety of designs. Materials of construction can vary; also humidity and possible freezing conditions have to be considered.

As a first step in shutdowns of cooling systems, the length of time involved must be examined. If the operation of a system is going to be interrupted only overnight or for a few days, the best procedure is to continue the operation of the recirculating pump without a heat load. Under no-load operation, it is necessary that nonfreezing conditions prevail. Continuing pumping will continue to provide a corrosion inhibitor to the metal surface and assist in preventing settling of sediment.

Should the shutdown time be lengthened to several weeks, the same procedure can be used, or the recirculating pump can be started about once a week for 1–2 h. This will provide fresh inhibitor to the metal surface and possibly move particulate matter that may have settled on the metal surface. Circulating the cooling water applies only to recirculating and closed systems, not once-thru systems. Water losses in once-thru units are much higher. In recirculating systems, the cooling tower fans should be idle to reduce water losses and also eliminate the energy costs of operating the fans.

If a shutdown is to be extended to months, more extensive steps must be taken. The steps selected must consider the size and complexity of the system. Should the climate be dry, the equipment can be drained. Where drainage is difficult, compressed air can be used to blow out remaining water.

When equipment can be sealed effectively, the dry methods of storage are effective. Item 1 below is an exception to the sealing requirement.

Dry methods of storage involve:

1. Leaving equipment open to allow free movement of air. This method is effective only in dry areas. Past operating history is the best guide for selecting a viable procedure.

2. Using a nitrogen blanket. Operating at 2–3 psig, the nitrogen displaces the oxygen present in air.

3. Using vapor *phase* inhibitors to protect tightly sealed units. The largest use of vapor phase inhibitors has

been to protect metallic goods in shipment; the air space is small. If the air space in cooling water equipment is not too large, the vapor phase inhibitors can be effective.

4. Using dessicants in closed areas, where they are effective in most instances. Equipment can range from trays of quick lime to commercial dessicant dehumidifiers.

Wet storage using high dosages of conventional corrosion inhibitors is effective when freezing conditions do not exist. If it can be used, chromate is the most effective product. Dosages can vary from 400–600 ppm CrO_4. At times, only a high pH (10–12) is required to prevent corrosion.

Without circulation, polyphosphates, zinc, and molybdates are limited in their effectiveness. In selecting an inhibitor, the type of metal in use has to be considered.

Where freezing conditions exist, inhibited ethylene glycol solutions should be added. Dosages vary with climatic temperatures.

For long scattered systems, it may not be economical to protect all equipment for long shutdown periods. It may be more practical to allow small-diameter pipe, etc., to corrode and to replace it as required.

Once started, corrosion pits are difficult to halt. For this reason, shutdown practices demand the same amount of diligent supervision as an operating unit.

11 Deposits

Cooling Water Deposits

Precipitation Deposits

The solubility of most salts increases with temperature. Important exceptions to this rule include calcium carbonate and calcium sulfate.

Calcium Carbonate. It has been estimated that calcium carbonate makes up one-fifth of the earth's crust. In the majority of cooling waters, its solubility has to be considered. The scaling tendencies of calcium carbonate are customarily calculated through the use of scaling indexes. The *Langelier, Ryznar,* and *Puckorius* indexes are the best known. The Langelier index (Table 11-1) is probably the most widely used. There are two main limitations to the use of this index. The tendency to form scale is indicated, but the amount is not. In addition, the index

Table 11-1. Data for Rapid Calculations of the Langelier Index (Calcium Carbonate Saturation Index)

A		C		D	
Total Solids (ppm)	A	Calcium Hardness (ppm of CaCO₃)	C	M. O. Alkalinity (ppm of CaCO₃)	D
50–300	0.1	10–11	0.6	10–11	1.0
400–1000	0.2	12–13	0.7	12–13	1.1
		14–17	0.8	14–17	1.2
		18–22	0.9	18–22	1.3
		23–27	1.0	23–27	1.4
Temperature (°F)	B	28–34	1.1	28–35	1.5
		35–43	1.2	36–44	1.6
		44–55	1.3	45–55	1.7
		56–69	1.4	56–69	1.8
		70–87	1.5	70–88	1.9
32–34	2.6	88–110	1.6	89–110	2.0
36–42	2.5	111–138	1.7	111–139	2.1
44–48	2.4	139–174	1.8	140–176	2.2
50–56	2.3	175–220	1.9	177–200	2.3
58–62	2.2	230–270	2.0	230–270	2.4
64–70	2.1	280–340	2.1	280–350	2.5
72–80	2.0	350–430	2.2	360–440	2.6
82–88	1.9	440–550	2.3	450–550	2.7
90–98	1.8	560–690	2.4	560–690	2.8
100–110	1.7	700–870	2.5	700–880	2.9
112–122	1.6	880–1000	2.6	890–1000	3.0
124–132	1.5				
134–146	1.4				
148–160	1.3				
162–178	1.2				

(1) Obtain values of A, B, C, and D from above table.
(2) $pH_s = (9.3 + A + B) - (C + D)$.
(3) Saturation index = $pH - pH_s$.
If index is 0, water is in chemical balance.
If index is a plus quantity, scale-forming tendencies are indicated.
If index is a minus quantity, corrosive tendencies are indicated.
Source: Based on the Langelier formula, Larson–Buswell residue, temperature adjustments; arranged by Nordell. Table 9.2 from *Water Treatment for Industrial and Other Uses* by Eskel Nordell, ©1961 by Reinhold Publishing. Reprinted by permission of Van Nostrand Reinhold Company, Inc.

is not efficient in indicating corrosive tendencies. The index is only one of many factors involved in predicting corrosive properties.

Calcium carbonate solubility is not influenced by increased temperatures only. Many ground waters contain an excess of carbon dioxide under pressure. This tends to depress the pH. When the water passes through a well screen, a pressure drop occurs, and the carbon dioxide is released. Calcium carbonate can then result even though there has been no temperature increase.

Concentrations of dissolved solids in open recirculating systems can result in calcium carbonate scale formation. The bulk of the scale forms, however, in heat exchangers, where the temperatures are elevated. Two tools are available to avoid calcium carbonate scale. Acid can be added to lower the pH and alkalinity; sulfuric acid is usually employed since it is the least expensive. The other method of control is to continuously add a scale inhibitor such as polyphosphate or phosphonate. These materials interfere with the crystalline growth of calcium carbonate.

Polyphosphates, because they present no toxicity problems, are commonly used in municipal and other potable systems. The reversion of polyphosphate to orthophosphate prevents its wider use in open recirculating systems. When polyphosphates are used in recirculating systems, they are usually employed not so much to prevent scale as for their corrosion-inhibiting features.

Phosphonates are now more widely used in open recirculating systems to control calcium carbonate scale. They are more effective than polyphosphates and are much more stable. Low-molecular-weight polyacrylates and sulfonated styrenes have also been used to control calcium carbonate deposits. Mixtures of scale inhibitors have been used with success.

If a scale sample contains a substantial amount of calcium carbonate, it will effervesce if a mineral acid like muriatic (hydrochloric) is added. For confirmation, a laboratory scale analysis should be made.

Calcium Sulfate.   In oil production, calcium sulfate scale is a serious problem. In industry, it is much easier to stay within the prescribed solubility limits. As a result, calcium sulfate scale problems are usually avoided. Much more accurate solubility studies have been made, but one general rule used in preventing calcium sulfate scale is

$$\text{ppm Ca}^{++} \times \text{ppm SO}_4^{=} = <500,000$$

Generally, in most systems, it is possible to concentrate a cooling water four to five times without exceeding the 500,000 figure. It is important to include the $SO_4^{=}$ added through the addition of sulfuric acid in making calculations. In a few cases, where the sulfate content is high, it may be necessary to substitute muriatic acid in place of sulfuric. The cost increase is substantial.

Some scale inhibitors, such as phosphonates and polyacrylates, are effective in preventing calcium sulfate scale. However, sulfate deposits are very difficult to remove. Accordingly, the recommended procedure is to stay within the solubility limit of calcium sulfate.

Silicates.   Silicates form a number of different scale complexes with calcium, magnesium, aluminum, sodium, and iron.

There is, at the present time, no effective dispersant for silicate deposits. The usual control procedure is to maintain the silica level in open recirculating water below 200 ppm. In most areas of this country, this presents no problem, and at least four or five cycles can be attained, based on silica content. In some southwestern areas, however, it is not unusual for raw-water silica levels to reach 60–80 ppm. This severely restricts the cycles of concentration that can be attained. Under these conditions, the best practice is to use a calcium carbonate scale-control agent and allow the pH to elevate to 8.0–8.5. The higher pH tends to keep the silicates in solution.

Magnesium Hydroxide.   Only occasionally are magnesium hydroxide deposits found in cooling water systems. A pH of

10.0^+ and increased temperatures are required for its formation. Polyphosphates are not effective in preventing its formation. The solution involves reducing both the pH and temperature.

Iron and Manganese Deposits

Iron and manganese deposits differ from calcium carbonate and sulfate in that the insoluble oxides do not necessarily form on the metal surface. They may form in the bulk solution. Also, precipitation results not from increased temperature but from the oxidation of the soluble ferrous and manganous salts to the insoluble ferric and manganic oxides. Where iron bacteria are involved, more oxide deposits can be expected in the wall slime area since higher levels of the oxidizing bacteria are found there than in the bulk solution.

Objectionable levels of iron and manganese are considered to be 0.3 ppm and 0.2 ppm, respectively. At or above these levels, treatment or removal is recommended.

Iron and manganese are best removed before they enter a cooling water system. In most cases, this requires considerable capital expenditure. Where iron and manganese have not been removed, they can often be controlled by adding sequestrants or dispersants. Iron deposits are reddish-brown, while manganese oxides are gray to black. Products usually added are polyphosphates or phosphonates. Polyphosphates are the automatic choice if the water is potable-grade. The recommended dosage is usually 2 ppm polyphosphate plus 2 ppm of polyphosphate per part of iron and manganese in the system. The feed of polyphosphate should be added before any contact with air or stronger oxidizing agents such as chlorine. Of the polyphosphates, sodium tripolyphosphate is the most effective, followed by sodium pyrophosphate and hexametaphosphate, in that order. High-hardness waters interfere with the performance of all polyphosphates in controlling dissolved iron and manganese. Since sodium hexametaphosphate is the most effective for controlling calcium hardness, the most effective product for controlling dissolved iron or manganese in a hard water is often a blend of two or more polyphosphates.

If the level of iron or manganese is above 2 ppm, phosphonates should be used instead of polyphosphates. They cannot be used if the water is potable-grade.

Corrosion Products

When corrosion occurs, deposits often form; the end result is reduced heat transfer as with other types of fouling.

The solution to this problem is to prevent corrosion. This has been covered in Chapters 9 and 10.

Oil

Oil is both the cause and result of fouling (Franco, 1984). It acts as a binder in adhering sediment to a metallic surface. As a component of the deposit, it also acts as insulation to heat transfer. When oil leaks occur, the source of the leaks should be determined, and the defective equipment component should be shut off from the main system until it is repaired. As much of the oil as possible should be removed by skimming operations. Two likely locations are the cooling tower deck and basin.

An initial dosage of surfactant (approximately 100 ppm) should be added to the system. Normal blowdown dosage should be maintained, as well as normal chlorination dosages, for one day. Successful oil removal with a nonionic surfactant (polysiloxane and polyalklene or equivalent) has been reported. The system should be blown down heavily for several days without addition of surfactant. Then, follow with surfactant addition for about one week at normal operation plus 100 ppm surfactant. Addition of surfactant often results in foaming. In such cases, a defoamer will have to be added.

This procedure can be modified as required, depending on the system and the amount of oil present.

Incompatible Water Treatment Chemicals

Deposits can develop when treatment chemicals are incompatible with each other or with the water in which they are used.

When anionic flocculants were first introduced to cooling water systems, they were often mixed with cationic biocides. Sticky deposits often resulted.

Carbamate biocides are often used in cooling water containing chromates. Reduction of the chromates follows, frequently resulting in chromium hydroxide deposits.

Problems often develop with the use of zinc inhibitors in high-pH waters. Insoluble zinc salts form in the cooling water and then settle on the metal surface. Under these conditions, no corrosion protection is provided.

All compatibility problems cannot be foreseen; but caution should be used when products with cation and anion charges are mixed. The same reasoning applies to oxidizing and reducing agents. Also, it should be determined whether the water treatment chemicals are soluble in the pH ranges that could be encountered. This applies when alkaline or acidic leaks can develop.

Polyphosphates are not compatible with high-calcium-hardness waters if a large amount of polyphosphate reverts to orthophosphate. Tricalcium phosphate deposits can develop. It is difficult to predict the reversion rate of polyphosphates; factors involved include temperature, time of residence, bacteria, and catalysts.

Sediment

Sediment in a cooling water can originate in the raw-water makeup, or it can be blown into a cooling tower. It is virtually impossible to predict all the types of sediment that can enter a system. Still, chances are excellent that much of the sediment will settle in low-velocity areas.

Sediment can be removed in side stream filters; also, equipment should be designed to avoid low-velocity surfaces. Operating personnel should make certain that velocities are not reduced to the danger zone through valve throttling. This is often done to control temperatures. Chemically, dispersants such as low-molecular-weight polyacrylates or sulfonated styrenes can be used to prevent the settling of sediment. The degree of success varies, depending on the type and amount of sediment.

The main defense against sediment forming in heat exchangers is to maintain sufficient velocity. A minimum velocity of 3 fps has been recommended to avoid sediment dropout. Design velocity should be somewhat higher, about 5 fps. Prevention of settling is more difficult on the shell side since baffles allow for many low-velocity areas.

Microbiological Fouling

Microbiological fouling (Figs. 9-1–9-3) differs from other types of deposits in that it involves living organisms. Slime deposits form rapidly, much faster than nonliving deposits. Scales such as calcium carbonate customarily develop at slow, predictable rates. Slime deposits, on the other hand, have been known to foul a heat exchanger in a week or two. If fouling develops in a fast, erratic manner, the chances are excellent that the fouling is microbiological in origin.

Dip-slide cultures have come into wide use in recent years as a monitor of biological growth. Although bacterial counts on dip slides represent the bulk water and not growth on the surface of the metal, they are an effective field tool.

An important factor in their field use is that the results of the sampling are available within 24–48 h. Before the introduction of the dip slides, water samples were customarily collected and mailed to a laboratory. It was not unusual for results to be delayed up to three weeks. The bacterial count results obtained were worthless.

In addition to the dip slides, the use of a microscope to examine deposits in the field is an excellent tool to determine if fouling is microbiological in origin; dip slides are not useful when microflora is composed primarily of algae and fungi.

These organisms are often implicated in biofouling. In many cases, the plant operator learns to correlate deposit problems with dip-slide bacterial counts and microscope examination.

Biocides—Limitations in Controlling Fouling

Indirectly, biocides contribute to deposit problems if they fail to control microorganisms involved in slime growth. Limitations of the most commonly used biocides are as follows:

Chlorine. The most widely used biocide is chlorine. Normally, in an open recirculating cooling water system, the chlorine is added to the cooling tower basin close to the inlet to the recirculating pump. If a particular heat exchanger is developing slime problems, the best control procedure is to add chlorine directly to the inlet to the heat exchanger.

Up to a pH of 8.0, chlorine is a strong oxidizing agent. In high-iron-and-manganese waters, chlorine will oxidize ferrous and manganous salts to the ferric and manganic form. Insoluble hydrated oxides will then develop. Dosages of chlorine required vary with individual systems. Ordinarily, a free-chlorine residual of 1 ppm for 1 h twice a day, when alternated with other biocides, will control most bacteria. Low dosages can result in loss of control. Chlorine is most effective at the low- and middle-pH range through the formation of hypochlorous acid. Effectiveness drops off rapidly above pH 7.5.

High-free-chlorine (> 1 ppm) residuals plus high pH are known to attack cooling tower wood, causing delignification. Delignification is characterized by a furry appearance on the surface of the wood. Little can be done to correct this damage. Accordingly, preventive measures should be taken; chlorine should be maintained at 1 ppm or lower in the cooling tower area, and pH should be maintained below 7.5. Although attack predominates in the flooded areas, it is also found in the air-inlet louvers and in the plenum area.

Another limitation of chlorine is its reaction with ammonia to form chloramines. Where ammonia leaks appear in a system, it is difficult to maintain any free chlorine. Chloramines have biocidal properties but are not considered as effective as free chlorine. Yet many investigators believe that there is a major kill of microorganisms by hypochlorous acid in the brief period before ammonia reacts to form chloramines or to be consumed by other chlorine demand.

Chlorine Dioxide. Although considerably more expensive than chlorine, chlorine dioxide is sometimes used in lieu of chlorine. Since chlorine dioxide, like chlorine, is a strong oxidizing agent, it will also oxidize ferrous and manganous waters. Unlike chlorine, it is effective over a wide pH range and also does not react with ammonia.

Low dosages will result in lack of biological control and slime buildup.

Quaternary Ammonium Compounds. As previously stated, where biocides are concerned, the chief problem regarding fouling is failure to perform. This is especially true for quaternary ammonium compounds in fouled systems.

Quaternary ammonium compounds are effective in maintaining a clean system. They are not suitable for fouled equipment. The "quats" tend to adsorb on virtually any surface. Slime includes not only dead and live bacteria but also dirt and other types of deposits. As a result, much of the quaternary ammonium biocide is wasted.

Carbamates. Organosulfur compounds, like carbamates, are effective, low-cost biocides. But their reducing properties must be recognized. They should not be used with chromate treatments since the hexavalent chromium will be reduced to trivalent chromium. Chromium hydroxide deposits will result in high-pH areas. Also, the loss of carbamate will allow slime to develop.

2.2 Dibromo-3-Nitrilopropionamide (DBNPA).   This oxidizing biocide has come into wide use in recent years. It decomposes at high-pH values. This limits its use to pH-controlled cooling water systems in the range of 5.5–7.5.

While a disadvantage in the cooling water system, DBNPA can be an advantage in disposal. Discharging into a higher-pH water results in its decomposition.

Methylene Bisthiocyanate.   This biocide has one of the same properties as DBNPA; that is, it decomposes in high-pH waters. pH control is required.

Cyanurates.   Cyanurates are chlorine donors. Since they have similar properties, the cyanurates should not be used in high-pH waters. Even though cyanurates are chlorine donors, they have been known to be more effective than chlorine gas in fouled recirculating systems. Increased stability and slow release may account for the increased effectiveness. A great deal of chlorine is lost in passing over a cooling tower; cyanurates do not have this limitation.

Organotin-Quats.   Comments on the limitations of quaternary ammonium compounds also apply to the organotin-quat compounds. Refer to Quaternary Ammonium Compounds, page 202.

Dodecylguanidine Hydrochloride.   This product has excellent dispersing properties. On the adverse side, it forms insoluble salts with high phosphates, sulfates, and chromates. Foaming can also be a problem.

Ozone.   An allotropic form of oxygen (O_2), ozone is an unstable blue gas with a pungent odor, and it readily decomposes to oxygen. Its solubility in water is low—about 10–15 times the solubility of oxygen.

Aside from fluorine, ozone is the strongest industrial oxidizing agent. As such, it is capable of removing iron, manganese, and other oxidizable contaminants. It is also a very effective biocide. Ozone disinfects by rupturing the cell wall. Unlike chlorine, ozone has limited residual effect because it decomposes to oxygen. The half-life of ozone in water at room temperature is about 18 min.

In recent years, there has been some use of ozone as a biocide in recirculating cooling water systems. Most successful applications have been in low-temperature, low-TDS waters. Air conditioning units predominate. In such applications, scale and corrosion problems are minimal; if bacterial problems are controlled, no other difficulties exist.

In most industrial recirculating cooling water systems, ozone has one serious drawback. High temperatures, plus scale-forming and corrosive waters, require the addition of water treatment chemicals in addition to the use of ozone. But ozone is such a reactive oxidizing agent that it often reacts with the treatment products. Under these conditions, ozone is not a satisfactory biocide.

Physical Deposit Factors

Surface Condition.   It has been established that smooth surfaces tend to foul to a lesser degree than rough surfaces. Corrosion tends to roughen a surface and promote deposit formation. For this reason, corrosion control is very important. Abrasive materials moving at high velocities also tend to promote a rough surface. Hence, the importance of using clean makeup water, free of suspended matter.

Velocity.   The influence of velocity in preventing deposits varies. Where sediment is concerned, high velocities incline to prevent settling on metallic surfaces. At no area of the cooling water system should the velocity be allowed to fall below 3 fps, if at all possible.

Where scale formation and microbiological fouling are concerned, high velocities often tend to increase deposit formation. The primary reason appears to be that a greater flow of
water contacts the surface; more scale-forming salts are available, and increased nutrients are provided for sessile microflora.

Temperature.   Scale formation is strongly influenced by temperature. Calcium carbonate, sulfate, and phosphate all exhibit
lower solubility at high temperatures.

Where bacteria are concerned, results will depend on the
optimum growing temperature for the particular organism.
Control will depend on the use of biocides.

Glossary

Acidic.   Referring to the amount of mineral acid and salts that hydrolyze to form hydrogen ions.

Aerobic.   Living in the presence of oxygen.

Algae.   Lower form of plant life; usually green because of chlorophyll.

Alkalinity.   The capacity of a water to neutralize acids. Common water alkalinities consist of bicarbonate, carbonates, hydroxide, phosphate, and silicate.

Aminomethylpropanol.   A type of neutralizing amine CH_3C $(CH_3)NH_2-CH_2OH$.

Amorphous.   Without a definite crystalline structure. Atoms or molecules are not arranged in a definite pattern.

Anaerobic.   Living in the absence of oxygen.

Anion.   A negatively charged ion of an electrolyte.

Anode.   Positive pole of a cell.

Austenitic.   A steel alloy that has a face-centered cubic lattice.

Bacteria.   Microscopic unicellular living organisms.

Bellows.   A metallic accordianlike box that can be compressed like a spring; used to accommodate expansion or contraction in piping systems, etc.

Biocide.   A toxicant that controls the growth of living organisms.

Buffer.   A solution that resists change in pH.

Carbonate.   A salt containing CO_3, such as sodium carbonate.

Carbonate Cycle.   A method of internal boiler water treatment. Calcium carbonate is precipitated in the presence of an organic polymer dispersing agent.

Carryover.   Moisture and entrained solids that carry over with steam.

Cation.   A positively charged ion of an electrolyte.

Caustic Cracking.   Cracking failure of metal caused by stress plus high concentrations of caustic.

Cavitation.   A type of corrosion caused by rapid change in pressure in corrosive environment; results from the formation and collapse of vapor bubbles.

Chelants.   Compounds that have the property of withdrawing metal ions into soluble complexes.

Coagulation.   Destabilization of colloids.

Color Throw.   The transfer of color from an ion-exchange resin to a liquid.

Condensate.   Water formed by condensation of steam. In general, any liquid formed by condensation of vapor.

Continuous Blowdown.   The continuous removal of concentrated boiler water to control dissolved and suspended solids.

Convection.   Heat transfer by fluid movement.

Coordinated Phosphate.   A type of buffer phosphate treatment used to prevent OH buildup in high-pressure boilers.

Corrosion Fatigue.   Effect of repeated fluctuating stresses in a corrosive environment.

Creep.   The slow heated deformation of metal under a smaller stress than would cause rapid rupture.

Crevice Corrosion.   Corrosion taking place where a specific area is isolated from the bulk of a solution: flange areas are one example.

Crystal (grain).   A volume of metal where the alignment of metals remains the same.

Cycles of Concentration.   Ratio of total dissolved solids in system cooling or boiler water system to makeup of feedwater TDS.

Cyclohexylamine.   A type of neutralizing amine, $C_6H_{11}NH_2$.

Deaerator.   A type of feedwater heater operating with water plus steam; designed to remove oxygen and free carbon dioxide from water.

Denitrifying Bacteria. The microbial breakdown of nitrates and nitrites to ammonia and nitrogen.

Density. Ratio of mass of a substance to a given volume.

Dezincification. The selective separation of zinc from copper alloys.

Diethylaminoethanol. A type of neutralizing amine $(C_2H_5)_2$ NCH_2CH_2OH.

Differential Aeration. The stimulation of corrosion at a localized area by differences in oxygen concentration in the electrolytic solution in contact with metal surfaces; frequently occurs under deposits.

Dispersant. A substance that tends to maintain solid particles in suspension.

Downcomer. Tube or pipe in a boiler that connects the headers. Water flows downward.

Dry Steam. Steam containing no moisture. Commercially dry steam contains not more than 0.5% moisture.

Economizer. A series of tubes located in path of flue gases; serves to reduce temperature of gases and preheats boiler feedwater.

EDTA. A type of chelating agent. Abbreviation for ethylene di-aminetetraacetic acid.

Effervescence. The rapid escape of gas from a liquid.

Embrittlement. Brittle intercrystalline corrosion cracking in highly stressed areas.

Enthalpy. A thermal property of a liquid defined as the stored mechanical potential energy plus internal energy.

Entropy. The theoretical measure of energy, as of steam, that cannot be transformed into mechanical work.

Erosion. The wearing away of metal by moving suspended matter or water.

Erosion-Corrosion. A corrosion reaction accelerated by the movement of the liquid against a metallic surface.

Exfoliation. Localized subsurface corrosion in zones parallel to surface. Corroded area can be removed in layers.

Failure. A rupture, break, or deterioration of metal.

Fatigue.   The fracture of a metal following repeated cycle stresses at below the tensile strength of the material.

Feedwater Heater.   A surface heat exchanger used to preheat feedwater with steam.

Ferrule.   A short tube rolled into tube hole; usually used to reduce erosion damage.

Filtration.   The separation of suspended matter from a liquid by passing the liquid through a fixed bed (e.g., sand) or filter paper.

Fines.   Small particles of ion-exchange resins.

Firing Rate.   The rate at which a fuel is supplied to a burner.

Fire Tube Boiler.   A boiler with water on the outside of straight tubes. Heating gases pass on inside.

Flocculation.   The collection of neutralized particles to form larger clumps, which settle.

Fouling.   A general term for deposits on surfaces.

Free Available Chlorine.   That portion of the total chlorine that has not reacted.

Galvanic Corrosion.   Corrosion resulting from the contact of dissimilar metals in an electrolyte.

General Corrosion.   Metal attack uniformly distributed over a surface.

Generating Tubes.   Boiler tubes used for the generation of steam.

Hardness.   A measure of the amount of calcium or magnesium present in water; expressed as parts per million (ppm) $CaCO_3$ or grains per gallon (gpg).

Hydrazine.   A nitrogen compound, N_2H_4, used to remove oxygen from boiler water.

Hydrogen Embrittlement.   A loss of ductility in metal caused by the absorption of hydrogen.

Hydroxyl.   The term used to describe the OH radical.

Humic.   Referring to organic matter in or on a soil; composed of partial or decomposed organic matter.

Hydroxyapatite.   A hydrated calcium phosphate; the type of phosphate deposit or sludge usually present in boilers.

Impingement.   Deterioration of metal caused by the striking of a moving substance against the metal.

Inhibitor.   A substance that reduces corrosion.

Intergranular Corrosion.   Corrosion that occurs at grain boundaries.

Ion.   A charged atom or radical that may be positive or negative.

Ion Exchange.   A reversible process by which ions are exchanged between solids and a liquid.

Iron Bacteria.   Bacteria that assimilate iron and excrete its compounds in their life process.

Langelier Index.   An index used to determine the scaling or corrosive tendencies of a water.

Latent Heat of Vaporization.   Heat given off by a vapor condensing or gained by liquid evaporation without a change in temperature.

Lime-Soda Softener.   Water softener employing calcium hydrate and sodium carbonate as the reacting chemicals.

Magnetite.   Black magnetic oxide of iron, Fe_3O_4.

Makeup Water.   Water that is added to a boiler or cooling water system to replace water lost.

Morpholine.   A type of neutralizing amine, C_4H_8ONH.

Nitrifying Bacteria.   Bacteria that oxidize ammonia and nitrites to nitrates.

Nitrogen Blanket.   A covering of nitrogen gas over water; used to prevent contact of oxygen with water.

NTA.   Abbreviation for chelating agent, nitrilotriacetic acid.

Oxide.   A compound composed of metal and oxygen.

Oxidizing Agent.   Substance that can take electrons from an atom or radical.

Organic Compound.   With a few exceptions, all compounds containing carbon.

Oxygen Attack.   Corrosion caused by oxygen.

Packaged Boiler.   A boiler equipped and shipped complete with firing equipment.

Pearlite.   Product that results when slowly cooled steel transforms at about 1328°F. It consists of alternate laminae of iron and iron carbide.

Phase.   One of the states of matter: liquid, gas, or solid.

pH.   Negative logarithm (base 10) of the hydrogen ion concentration.

Phosphate.   The PO_4 radical.

Phosphonates.   Derivatives of the hypothetical phosphoric acid $HP(O) (OH)_2$.

Pitting Corrosion.   Localized corrosion, forming cavities.

Polyacrylate.   Synthetic polymer formed from the monomer, acrylic acid $H_2C:CHCOOH$; used as a boiler and cooling water dispersing agent.

Polymers.   Large organic molecules resulting from the joining of single monomers.

Polyphosphate.   Molecularly dehydrated orthophosphate.

Pressure Gauge.   Measurement of pressure above atmospheric.

Radiant.   The property that permits heat to be transferred by rays. To receive rays, the object must be in direct path.

Radiant Superheater.   A superheater exposed to the direct rays of the fire.

Raw Water.   Untreated water.

Reducing Agent.   Substance that gives up electrons.

Regeneration.   Restoration of water treating ability of an ion-exchange resin.

Reverse Osmosis.   A selective membrane method of separating dissolved salts from water.

Riser Tube.   A tube through which steam and water rise in a water tube boiler.

Ryznar Index.   An index to determine the scaling or corrosion tendencies of a water.

Scale.   A deposit of medium to extreme hardness occurring in both boiler and cooling systems.

Slime.   A soft, sticky mucuslike substance originating from bacterial growth.

Sludge.   A soft mud.

Steam Blanket.   A covering of steam over water.

Steam—Dry Saturated.   Steam at the saturation temperature corresponding to the pressure and containing no water in suspension.

Steam Quality.   The percentage of vapor by weight present in a steam and water mixture.

Steam—Wet Saturated.   Steam containing water particles.

Stress Corrosion Cracking.   Crack corrosion caused by tensile stress and a specific corrosive environment.

Sulfate.   A salt containing the SO_4 radical.

Sulfate-Reducing Bacteria.   Bacteria that assimilate oxygen from sulfate compounds, reducing them to sulfide.

Superheated Steam.   Steam at a temperature higher than the saturation temperature corresponding to the pressure.

Superheater.   A heat exchanger in which additional heat is added to saturated steam.

Surfactant.   A surface active agent.

Suspended Solids.   Undissolved solids suspended in water.

Thermal Shock.   A stress strain condition set up by a sudden change in temperature.

Transgranular Corrosion.   Corrosion that occurs without regard for grain boundaries.

Water Hammer.   A sudden increase in pressure of water due to the instantaneous conversion of momentum to pressure.

Water Tube Boiler.   A boiler with water on the inside of tubes—heated on the outside.

Zeolite Softener.   A loosely used term to designate a sodium-regenerated ion-exchange softener; originally referred to a type of ion-exchange resin.

References

ASME Committee on Failure of Boilers and Pressure Vessels, "Failure of Boilers and Related Steam-Power Plant Equipment," ASME Report, New York, 1972.

Barer, R. D., and Peters, B. F., *Why Metals Fail*, Gordon and Breach, New York, 1970, p. 50.

Bell, A. W., and Breen, B. P., "Converting Gas Boilers to Oil and Coal," *Chemical Engineering*, April 26, 1977, pp. 93–101.

Bernard, P., "Tighten Up Condenser Operation by Checking Its Air Removal System," *Power*, Dec. 1981, pp. 30–31.

Bosich, J. F., *Corrosion Prevention for Practicing Engineers*, Barnes and Noble, New York, 1970.

Bumbard, R. J., "Troubleshooting Condensate Systems," *Power Engineering*, Aug. 1982, pp. 65–68.

Butler, G., and Ison, H. C., *Corrosion and Its Prevention in Waters*, Reinhold, New York, 1966, p. 157.

Castagna, C. J., and Miller, W. S., "Understanding Ion-Exchange Resins for Water Treatment Systems," *Plant Engineering*, March 19, 1981, pp. 191–193.

Chemical Engineering Staff, Discussion, "Update on Reverse Osmosis," *Chemical Engineering*, July 9, 1973, p. 8.

Crandall, H. E., "Test Methods for Biocide Evaluation in Cooling Towers," *Journal of the Cooling Tower Institute*, Vol. 5, No. 1, 1984, pp. 9–14.

Cuisa, D. G., "New Trends in Boiler Scale Control," paper presented at International Water Conference, Pittsburgh, 1973.

Davies, V. R., "Organic Contamination of Source Waters and Its Control in Relation to Fouling of Anion Exchange Resins in Demineralization Plants, *Combustion*, May 1979, pp. 40–45.

Davies, V. R., "Operation and Maintenance of Ion-Exchange Equipment," *Corrosion and Maintenance*, July–Sept. 1984, pp. 185–195.

Duolite Staff, "Ion Exchange in Water Treatment," Duolite International, Inc., Redwood City, Calif., 1982.

Dupree, N. E., "Economics of Feedwater by Reverse Osmosis," *Industrial Engineering*, May–June 1973, pp. 36–48.

Evans, U. R., *The Corrosion and Oxidation of Metals*, Edward Arnold, London, 1960, p. 707.

Firman, E. C., "Waterside Flow Problems in Boilers," *Combustion*, July 1975, pp. 21–28.

Fontana, M. G., and Greens, N. D., *Corrosion Engineering*, McGraw-Hill, New York, 1967, p. 39.

Franco, R. J., "Minimizing the Deleterious Effects of Oil Leakage in Cooling Water Systems," *Materials Performance*, March 1984, pp. 23–27.

French, D. W., *Metallurgical Failures in Fossil Fired Boilers*, Wiley, New York, 1983.

Frey, D. A., "Case Histories of Corrosion in Industrial Boilers," *Materials Performance*, Feb. 1981, pp. 49–56.

Gelosa, L. R., and McCarthy, J. W., "Latest Chemical Treatment," *Power*, Jan. 1979, pp. 78–81.

Godard, H. P., Jepson, W. B., Bothwell, M. R., and Kane, R. L., *The Corrosion of Light Metals*, Wiley, New York, 1967.

Good, R. B., "Performance of HEDP in Boilers," *Materials Performance*, Sept. 1961, pp. 29–32.

Hamer, P., Jackson, J., and Thurston, E. F., *Industrial Water Treatment Practice*, Butterworths, London, 1961.

Holmes, D. R., and Mann, G. M., "A Critical Study of Possible Factors Contributing to Internal Boiler Corrosion," *Corrosion*, Nov. 1965, pp. 371–376.

Hwa, C. M., Bodach, C. M., and Schroeder, C. D., U.S. Patent 3,666,404, "Method of Inhibiting Corrosion in Aqueous Systems with High Molecular Weight Alkylene Oxide Polymers," 1972.

Industrial Water Engineering Staff Report, "RO: A Profile," *Industrial Water Engineering*, May–June 1973, p. 8.

Kaup, E. C., "Design Factors in Reverse Osmosis," *Chemical Engineering* , April 2, 1973, pp. 47–54.

Keys, L. H., "The Corrosion of Stainless Steels," *Australian Corrosion Engineering*, March 1976, pp. 9–16.

Kirby, G. W., "Corrosion Performance of Carbon Steel," *Chemical Engineering*, March 12, 1979, pp. 72–84.

Klein, H. A., "Corrosion of Fossil Fueled Steam Generator," *Combustion*, Jan. 1973, p. 13.

Kosarels, L. J., "Purifying Water by Reverse Osmosis," *Plant Engineering*, Aug. 9, 1979, pp. 183–185.

Kremen, S. S., "Reverse Osmosis for Water and Waste Treatment," paper presented at International Water Conference, Pittsburgh, 1970.

Kunin, R., "Helpful Hints in Ion Exchange Technology," reprinted from *Amber HI-Lites*, Rohm and Haas, Philadelphia, 1962.

Leferre, L. J., "Handling Ion Exchange Systems for Top Performance," *Power*, Aug. 1982, pp. 51–54.

Limpert, G. H., and Schroeder, C. D., "Sulfide Deposits in Sulfite Treated Boilers," *Heating/Piping/Air Conditioning*, Nov. 1978, pp. 89–91.

Lindinger, R. J., and Curran, R. M., "Experience with Stress Corrosion Cracking in Large Steam Turbines," *Materials Performance*, Feb. 1982, pp. 22–26.

Lutey, R., "Microbiological Corrosion," International Corrosion Forum, Chicago, 1980.

Marshall, A., "The Inhibition of Ferrous Metal Corrosion in Cooling Water Systems by Organophosphorous/Zinc Formulations," Corrosion 81 Forum, sponsored by NACE, Toronto, 1981, Paper 192.

McCoy, J. W., *The Chemical Treatment of Boiler Water*, Chemical Publishing, New York, 1981.

Mitchell, K. E., and Schroeder, C. D., "How to Prevent Deposits on Valve Stems and Turbine Blades," *Power Engineering*, Dec. 1964.

Monroe, E. S. Jr., "Effects of CO_2 in Steam Systems," *Chemical Engineering*, March 1983, pp. 209–212.

Moore, R. E., "Selecting Materials to Resist Corrosive Conditions—Part 2," *Chemical Engineering*, July 30, 1979.

Mulloy, G. Sr., Nashville Chemical & Equipment Co., Inc., Nashville, Tenn. Correspondence, 1989.

Murphy, V. P., "Chemical Cleaning Industrial Steam Generators," *Combustion*, May 1979, pp. 40–45.

NACE Task Group T-7G-1 Report, "Investigation of the Effects of Corrosion Inhibiting Treatments on Mechanical Seals in Recirculating Hot Water Systems," *Materials Performance*, July 1981, pp. 53–58.

Nachod, F. C., and Schubert, J., *Ion Exchange Technology*, Academic Press, New York, 1956.

Nathan, C. C., *Corrosion Inhibitors*, NACE, Houston, 1973, p. 143.

O'Keefe, W. O., "Condenser Venting System Gets Attention in Power Plant's Conservation Drive," *Power*, Dec. 1973, pp. 49–52.

Owens, D. L., *Practical Principles of Ion Exchange Water Treatment*, Tall Oaks Publishing, Littleton, Colo., 1985, p. 35.

Peters, C. R., "Fundamentals of Demineralization," paper presented at Liberty Bell Corrosion Course, Sept. 23, 1982.

Peters, C. R., "Water Treatment for Industrial Boiler Systems," *Industrial Water Engineering*, Nov.–Dec. 1980.

Schroeder, C. D., "Corrosion in Cooling Water Systems," *Australian Chemical Engineering*, June 1972, pp. 15–18.

Schroeder, C. D., "Water Treatment Case Histories—The Nonconformists," *Corrosion and Maintenance*, July–Sept. 1982, pp. 209–211.

Schroeder, C. D., and Landborg, R. J., "Understanding pH Oddities in Water Treatment," *Plant Engineering*, Oct. 14, 1976, pp. 155–156.

Schwartz, M. P., "Four Types of Heat Exchanger Failures," *Plant Engineering*, Dec. 23, 1982, pp. 45–50.

Shields, C., *Boilers*, McGraw-Hill, New York, 1961.

Sohre, J. S., "Causes and Cures for Silica Deposits in Steam Turbines," *Hydrocarbon Processing*, Dec. 1972, pp. 87–89.

Stephans, J., and Walker, J., "Predicting Chelate Performance in Boilers," *Industrial Water Engineering*, July–Aug. 1973, pp. 30–32.

Syrett, B. C., and Colts, R. L., "Causes and Prevention of Power Plant Condenser Tube Failures," *Materials Performance*, Feb. 1943, pp. 44–50.

Tatnall, R. E., "Fundamentals of Bacteria Induced Corrosion," *Materials Performance*, Sept. 1981, pp. 32–38.

Trace, W. L., "Condensate Corrosion Inhibition—A Novel Approach," International Corrosion Forum, Chicago, March 1980.

Trigs, L. E., "Improving Boiler Reliability," *Combustion*, Oct. 1978, pp. 33–60.

Tuthill, A. H., "Sedimentation in Condensers and Heat Exchangers: Causes and Effects," *Power Engineering*, June 1985, pp. 46–49.

Uhlig, H. H., *Corrosion and Corrosion Control*, Wiley, New York, 1971.

Veerabhadra Rao, P., Seetharaniah, K., and Rama Chav, T. L., "Cavitation Corrosion and Its Prevention by Inhibitors and Cathodic Protection," *Corrosion Prevention and Control*, Dec. 1972, pp. 8–13.

Wilson, R. A., "What Causes Corrosion in the Condenser Steam Space?", *Power Engineering*, Reprint, 1961.

Index